Adolescent Gynecology

A GUIDE FOR CLINICIANS

Adolescent Gynecology

A GUIDE FOR CLINICIANS

Edited by

Alfred M. Bongiovanni, M.D.

University of Pennsylvania School of Medicine
Pennsylvania Hospital
Philadelphia, Pennsylvania

PLENUM MEDICAL BOOK COMPANY
New York and London

Library of Congress Cataloging in Publication Data

Main entry under title:

Adolescent gynecology.

Includes bibliographical references and index.
1. Pediatric gynecology. 2. Youth—Sexual behavior. 3. Pregnancy, Adolescent. I.
Bongiovanni, Alfred M., 1921–   . [DNLM: 1. Genital diseases, Female—In ado-
lescence. WS 360 A2393]
RJ478.A35 1983   618   82-22396
ISBN-13:978-1-4615-8326-4   e-ISBN-13:978-1-4615-8324-0
DOI: 10.1007/978-1-4615-8324-0

© 1983 Plenum Publishing Corporation
233 Spring Street, New York, N.Y. 10013
Softcover reprint of the hardcover 1st edition 1983

Contributors

Dale M. Allison, B.S.N., Department of Obstetrics and Gynecology, School of Medicine, University of Pennsylvania, Philadelphia, Pennsylvania 19104

Gregory C. Bolton, M.D., Department of Obstetrics and Gynecology, Pennsylvania Hospital, Philadelphia, Pennsylvania 19107

Alfred M. Bongiovanni, M.D., Departments of Pediatrics and Pediatrics in Obstetrics and Gynecology, School of Medicine, University of Pennsylvania, Philadelphia, Pennsylvania 19104; Department of Perinatal Endocrinology, Pennsylvania Hospital, Philadelphia, Pennsylvania 19107; and Department of Pediatrics, Thomas Jefferson University, Philadelphia, Pennsylvania 19107

Elisa Bongiovanni, J.D., M.A., Thomas Jefferson University, Philadelphia, Pennsylvania 19107

Helen O. Dickens, M.D., Department of Obstetrics and Gynecology, School of Medicine, University of Pennsylvania, Philadelphia, Pennsylvania 19104

James D. Garnet, M.D., Department of Obstetrics and Gynecology, Pennsylvania Hospital, Philadelphia, Pennsylvania 19107

Alvin F. Goldfarb, M.D., Department of Obstetrics and Gynecology, Pennsylvania Hospital, Philadelphia, Pennsylvania 19107

George R. Huggins, M.D., Department of Obstetrics and Gynecology, University of Pennsylvania Hospital, Philadelphia, Pennsylvania 19104

Leonore C. Huppert, M.D., Department of Obstetrics and Gynecology, Medical College of Pennsylvania, Philadelphia, Pennsylvania 19129

Harold I. Lief, M.D., Department of Psychiatry, Pennsylvania Hospital, Philadelphia, Pennsylvania 19106; and School of Medicine, University of Pennsylvania, Philadelphia, Pennsylvania 19104

Edward E. Wallach, M.D., Department of Obstetrics and Gynecology, School of Medicine, University of Pennsylvania, Philadelphia, Pennsylvania 19014; and Department of Obstetrics and Gynecology, Pennsylvania Hospital, Philadelphia, Pennsylvania 19107

Joseph A. Zeccardi, M.D., Department of Emergency Medicine, Division of the Department of Surgery, Thomas Jefferson University Hospital, Philadelphia, Pennsylvania 19107

Foreword

ABOUT THE SUBJECT MATTER

Adolescence is a time of significant change. The adolescent era spans the interval between childhood and adulthood. It is a time of physical, social, and even emotional upheaval. During this relatively brief period of time not only does physical growth accelerate but, of more significance in the biology of any species, the individual attains reproductive maturity. Within this time frame, the human being acquires the capacity to procreate and perpetuate our species. From a reproductive standpoint, throughout adeolescence the individual is being prepared for perhaps what might be considered the most important function an organism has during his/her brief sojourn on earth, namely to endow successor(s) with a minute quantity of DNA to enable continuation of its form of life. The interlude between childhood and adulthood is not always socially or emotionally simple. The anatomic and physiologic modifications that come to pass during adolescence are not necessarily paralleled by a capacity to assume the societal responsibilities associated with reproductive maturity. Although the physiologic changes normally proceed in a predetermined fashion, adaptation to a changing role is a far more trying process.

Accustomed to living in a child's world, the subject of adolescent change requires considerable time and understanding to facilitate existence in the adult world. The early stages of sexual awareness, for example, are generally extremely confusing and, as evidenced by the inordinately high frequency of teenage pregnancy and sexually transmitted diseases, can be extraordinarily dangerous. Furthermore, when adolescence is complicated by disorders such as delayed or accelerated puberty and chromosomal defects leading to difficulties in sexual identification, the usual problems in adjustment are compounded. The adolescent female not only matures earlier than her male counterpart, but also encounters significantly more changes in the process. Dramatic changes in body size and shape, initiation of menstruation, and awareness of her vulnerability to pregnancy contrast strikingly to the relatively few alterations occurring at the comparable time in her male counterpart.

Only months earlier the adolescent female had fallen under the medical aegis of the pediatrician. Now who should be identified as the physician responsible for her care? One alternative is the pediatrician, whose time is absorbed by problems related to infant nutrition, infectious diseases of childhood, and immunization schedules. Another option is the gynecologist, but his waiting room is occupied by expectant mothers, women undergoing annual screening for reproductive tract disorders, and patients seeking contraceptive care. Does the adolescent female fall readily into one of these two camps? Perhaps, the medical needs of the adolescent female reflect the staggering series of physical and physiologic changes to which she is being exposed, often without the preparedness to handle them. To whom does she turn for medical care during this rapidly changing era of her life? During this critical interval, the groundwork for her entire future existence is being established. The need for appropriate attention and guidance are obvious. The adolescent must be prepared to enter adulthood with the confidence of normal sexual development, prepared to meet the challenges of reproductive maturity with sound judgment, and prepared to comprehend the physical, anatomic, and emotional manifestations of her adolescence.

ABOUT THE EDITOR

Alfred Bongiovanni has provided a unique text. He has recognized the needs of the female adolescent as well as the needs of the medical profession for guidance in the area of adolescent gynecology. The editor is one of the most distinguished pediatricians of our time. His versatility is apparent in having served as Physician-in-Chief of Childrens Hospital of Philadelphia and Chairman of the Department of Pediatrics at the School of Medicine, University of Pennsylvania, while representing one of a small group of pioneers in the field of endocrinology of growth and development. Indeed, Dr. Bongiovanni is widely recognized for elucidation of the various enzymatic disorders in adrenal steroid biosynthesis. Having distinguished himself as a medical school dean, a respected researcher, and currently a professor in a Department of Obstetrics and Gynecology, our editor nonetheless remains a consummate clinician. Few educators are as qualified as Dr. Bongiovanni to recognize the needs of the adolescent female and to develop a text which faciliates fulfillment of these needs.

ABOUT THE READER

The reader may be a pediatrician, a family physician, internist, obstetrician/gynecologist, or perhaps even a psychologist or counselor. Each of these disciplines has developed, through trial and error, some appreciation of the problems unique to the adolescent. Unfortunately few of us are totally endowed with the skills required to handle the enormous spectrum of problems to which the adolescent female is exposed. One might predict that if Adolescent Obstetrics and Gynecology were to be recognized as a specialty in its own right, the volume of work and degree of challenge for each member of this specialty would be limitless. Since such a specialty does not exist at present and perhaps might not even be a forthcoming possibility, the readers, originating from diverse disciplines, will find themselves more prepared to deal with the adolescent female and her problems by virtue of exposure to the principles and practicalities provided in this text.

Edward E. Wallach

Preface

It is important that the provider of health care for the adolescent girl have an appreciation of the normal physiology and the pathology of this critical stage in development. The text was designed with several goals. The presentation of the biologic background and the physical manifestations of normal puberty are to serve as the framework against which deviations may be perceived. The laboratory guidelines in terms of hormonal changes are detailed insofar as possible. Sexual behavior in contemporary society and its consequences are set forth in several chapters. The early manifestations of dysfunction, which may affect normal development and subsequent fertility, are covered in the presentation of pubertal development and ovarian dysfunction. With the advances in both preventive and therapeutic oncology, it was important to include a discussion of tumors. Finally, with the complications of legal elements in dealing with the young, which vary with time and with place, this matter warranted discussion. It is hoped that this volume will be useful to practitioners within the medical and allied professions who are involved in the management of the young girl.

I feel that each topic has been prepared by highly experienced authorities. I acknowledge with great pleasure the inspiration and guidance of Dr. Edward E. Wallach in the preparation of this work. He is also grateful for the secretarial and the editorial assistance of Mrs. Celia A. Mellon and Miss Anne T. Barrett.

Alfred M. Bongiovanni, M.D.

Philadelphia

Contents

Chapter 5
Teenage Pregnancy
Helen O. Dickens and Dale M. Allison

Chapter 6
Contraceptive Use among Adolescents
George R. Huggins

Chapter 7
Venereal Disease in Adolescents
Gregory C. Bolton

Chapter 8
Rape of the Adolescent
Joseph A. Zeccardi

CHAPTER 1

The Initial Encounter

Alvin F. Goldfarb

1. INTRODUCTION

The adolescent patient poses special problems in the delivery of health services to women. These problems are attitudinal on the part of the practicing physician, the feelings of the adolescent herself, and the environment in which the program of health care is delivered. Most offices are not set up or designed for the adolescent patient. If possible, special hours should be set aside for the management of the adolescent female. It is not conducive to a comfortable relationship if the youngster must wait her turn in a reception room crowded with older women many of whom may be pregnant. Much of the conversation which occurs in this setting is inappropriate and sometimes distressing. Although it is not always advisable, it is helpful if she be accompanied by a parent, preferably the mother or a close relative.

The typical adolescent is understandably anxious and concerned regarding any examination, particularly a gynecologic exam. She fears the possible discovery of an abnormality or serious illness. She expresses concern over the interview and the vaginal examination, considering it an invasion of privacy. In many instances, she has not been properly prepared for the examination and is anxious about the possibility of having to pay for the office visit. The physician's intelligent and empathetic approach to this patient is most important in providing care appropriate to the specific problem.

It is essential for the physician to establish a rapport with the young patients. This means one must respect the individual, maintain confidentiality, and be a friend. One may not always agree with her thoughts or behavior but must keep in mind her stage of emotional development. The

Alvin F. Goldfarb • Department of Obstetrics and Gynecology, Pennsylvania Hospital, Philadelphia, Pennsylvania 19107.

physician must endeavor to give the patient guidance by helping her to evaluate herself and her own world. Since the adolescent is particularly sensitive to body image, sexuality, and sex structures, she requires a careful examination, a thorough history, and a measure of reassurance regarding gynecologic matters to avoid embarrassment in the development of a distorted self-image.

The gynecologic examination should be used as an educational experience for the individual. It is during this first encounter with the gynecologist that the adolescent can be introduced to meaningful information concerning sexuality, contraception, pregnancy, venereal disease, and the concept of health maintenance examinations throughout life.

2. THE PHYSICIAN

The physician in seeing the adolescent must understand that the approach to care must be impartial. He or she must be able to gain the patient's confidence initially and obtain a meaningful history. Most adolescents are interested in talking about only the present illness. Therefore, it is essential that the past medical history be thorough. Statements concerning the mother's pregnancy, the patient's growth and development, medical and surgical illnesses, and the onset of sexual maturation should be well documented. The adolescent is more often than not uninformed about familial medical history as well as specifics regarding past personal illness or hospitalization. The physician must also understand that approximately 50% of the adolescents between the ages of 15 and 19 are sexually active. Half of these young people do not use contraception and few use contraception at the first sexual experience. More specifically, studies show that between 1 and 3 million single female teenagers in the United States need contraception, therefore contraceptive counseling and services are important. These matters are further amplified in Chapter 4.

It is difficult for the physician to be nonjudgmental about the adolescent's sexuality. One might approach that part of the interview in the following fashion: First, ask the patient if there are any questions she might want to ask concerning contraception, venereal disease, or pregnancy. This will allow her to answer either yes or no. Should the answer be no, then proceed to give information about the need to understand contraception, the problems of venereal disease, and the problems of an unwanted pregnancy. Presenting the adolescent with information on these subjects will usually raise questions in her mind. At this time, she may share her feelings about these problems.

If, on the other hand, she answers the question in the affirmative, one should directly ask about her form of contraception. If none is used, a dialogue concerning its advisability is developed. Again, the problems of unwanted pregnancy and venereal disease are reviewed.

This approach avoids having the adolescent feel put upon. It is not advisable to ask directly, "Are you sexually active?" Such a direct question would offend the patient and probably arouse suspicions concerning the physician. It is acceptable to represent a parent figure until the patient is comfortable. If the sexually active adolescent is unwilling to share information with her parents concerning this matter, she is often unable to do so with her physician. However, at the end of the interview and visit, it may be helpful to ask the patient, "What would you like me to share with your parents concerning your visit?"

Questions about contraception are usually more directly answered by the adolescent. Once she indicates that she is sexually active, she will reveal the form of contraception that she uses and the adequacy of its employment. Most adolescents below the age of 16 fail to use any method properly.

In obtaining the history, the physician must be comfortable understanding sexual maturation both from the behavioral and the endocrine standpoint. These matters are considered elsewhere, but the physician should be cognizant of the importance this information has in the management of the adolescent patient.

In obtaining the history, one must be impartial, sympathetic, trustworthy, and above all demonstrate respect for the patient as an individual.

3. THE PATIENT

The adolescent patient is especially sensitive to long waits in crowded clinics at distances from home and to perfunctory disinterested inquiries, frequent changes of the staff, and lack of privacy. Such conditions will deter her from continuing care. Whenever possible there should be separate facilities for adolescent health care delivered by a staff sympathetic to the adolescent.

The usual problems prompting gynecologic care of the adolescent include: retarded or accelerated sexual development, menstrual dysfunction, pelvic pain, vaginitis or vulvitis, pregnancy, contraception, sexual counseling, or evaluation of possible gynecologic neoplasia. Each of these possibilities raises concerns about underlying disease, an abnormality in female development, or sexual malfunction. These concerns are in addition to a basic fear of intrusion of privacy and anxiety over the pain associated with an examination.

The patient is for the first time sharing her feelings with an adult who may or may not represent father or mother. Most adolescents, however, are forthright, honest, and nonmanipulative. However, they still act as other patients who have perceptions of what the physician wants in the way of answers. Therefore, in interviewing the adolescent, one must be indirect. In talking to the adolescent patient, it is recommended that one not use their language which is for their peers. Questioning should be in simple lay language. If one uses words that are of the adolescent style, the patient will become uncomfortable and feel "put down."

4. THE PHYSICAL EXAMINATION

In examining an adolescent for gynecologic purposes, it is important that the patient be placed at ease, the surroundings in which the examination is performed be pleasant and comfortable, and that the examination be done in such a manner that it becomes an educational experience for the patient. In doing this and in preparing the adolescent for a gynecologic examination, the physician should give an explanation of each step of the examination. It is not always necessary to perform a traditional gynecologic examination. This is especially true if the girl is not sexually active or has no specific complaints related to the internal genitalia. A bimanual rectoabdominal examination often suffices. With experience, this can be revealing with regard to palpation of the uterus and adnexa.

To examine the adolescent one should have: (1) a small specula of either the Peterson or narrow blade type or the typical Graves speculum or duckbill type; (2) cotton-tipped applicators available to obtain material from the vaginal wall to study under the microscope; (3) a well-lit room; (4) the adolescent properly draped and comfortable; (5) a nurse or female assistant; (6) visual aids to depict the examination.

At the time of the examination, the physician should explain: (1) why the examination is being performed; (2) what information will be obtained from the examination; (3) the anatomy of the reproductive system so that the adolescent may learn more about her own body.

In addition to the pelvic examination, the adolescent should undergo a breast examination. There is much discussion as to whether adolescents should be taught self-breast examinations, because of the problems with self-image. Some authors suggest that the adolescent may become too fixed on this particular portion of the anatomy. We usually examine the breasts and point out to the adolescent that it is our responsibility to examine them on an annual basis for medical disease, and there is no reason for her to do it continuously.

When one performs a pelvic examination, the use of mirrors to demonstrate the appearance of the external genitalia may be used as a technique for demonstrating the anatomy. The bimanual examination should be carried out as gently as possible. If indicated, material can be obtained from the posterior fornix for microscopic evaluation. The need for Papanicolaou smears should be based upon whether the patient is sexually active, has been pregnant, or is using oral contraceptives and when the pelvic examination reveals cervical erosion or eversion.

In most instances, the general physician should be able to recognize a normal epithelialized cervix and obtain routine "Pap smears." The patient exposed to diethytshbestrol during intrauterine life should be referred to a gynecologist competent in the use of the colposcope for further evaluation of the vagina and cervix and the detection of adenosis. All examinations must include a general physical examination.

5. THE CONSULTATION

If one is to obtain the respect and confidence of the adolescent, it is necessary that the history be obtained and the examination be performed in privacy. One need not routinely involve parents in these first two steps. After the examination and the dialogue in consultation, one should ask for permission to involve the parents in the discussion. The physician should explain the plans for treatment with the parents and ask permission for doing the same. This step is essential in the management of an adolescent problem. It is well to protect the adolescent from exposures which make her uncomfortable. Most adolescents do not want a discussion of contraception, venereal disease, or pregnancy in the presence of their parents. However, it is essential, should an adolescent be pregnant, that her parents be involved in the discussion of the medical management. This is a crisis in the life of most adolescents, and it is essential that they share this information with their parents so that the support of the family unit can be properly utilized in guiding her through the crisis.

Although many teenagers feel that their mothers do not know about their sexual activity, it is sometimes helpful to explain to an adolescent that her mother is also a sexual being and she probably suspects that her daughter is engaging in sexual relationships. Inform the patient that her mother may be relieved to know that her daughter has found a gynecologist, is seeking care, and is behaving in a responsible manner. If the adolescent still insists that her mother not be told, her wishes should be respected. The legal aspects are considered in Chapter 11.

The entire process should be an educational experience for the adolescent. Contraception is reviewed as well as problems of adolescent obstetrics, venereal disease, nutrition, and the need for health maintenance examinations. Self-respect and responsibility in human sexuality are discussed.

In the past decade, there has been an epidemic of pregnancies in adolescence. This problem is formidable. Greater sexual activity alone does not account for the substantially greater number of unwanted pregnancies among adolescents. Ignorance, irresponsibility, and immature attitudes are common in this age group.

6. THE TEAM APPROACH

In some institutions the adolescent is seen by a multidisciplinary team consisting of the physician, the psychiatric social worker, and health educators. In private practice, the team approach is not cost-effective. It is recommended, however, that one member of the physician's team in the private office should be skilled in talking with the adolescent about problems relating to general health. This individual can also serve as a counselor to the adolescent.

7. SUMMARY

It is essential for the physician to establish a rapport with the young patient. This entails a demonstration of respect for the patient as an individual, an assurance of confidentiality, and a friendly demeanor. Since the adolescent is particularly sensitive to body image, sexuality, and sex structures, she requires a careful explanation and a measure of reassurance in order to avoid the development of a distorted self-image.

The adolescent patient requires not only routine gynecologic care but general health care. The evaluation and treatment of gynecologic conditions in this age group should include an understanding of menstrual physiology, principles of contraception, obstetrics, and an awareness of the behavioral aspects of adolescent care.

CHAPTER 2

Pubertal Development

Edward E. Wallach
and Alfred M. Bongiovanni

1. INTRODUCTION

The germ cell complement in the human female is established prior to the time of her birth. This characteristic contrasts to germ cell production in the male. Whereas in the male, germ cells are continually produced throughout reproductive life, in the female, the pool of germ cells designated within the ovaries at birth undergoes progressive depletion until reproductive capacity concludes at the time of menopause. Because of this progressive decline in germ cell complement, termination of reproductive potential in the female customarily occurs at an earlier age than in the male. This basic difference in reproductive potential between male and female in the human requires review of gonadal embryology and ovarian physiology through that point at which the full range of ovarian functions has been established.

2. OVARIAN EMBRYOLOGY

The ovary and testis each arise from three sources: (1) the coelomic epithelium, (2) the subjacent mesenchyme of the mesonephric ridge, (3)

Edward E. Wallach • Department of Obstetrics and Gynecology, School of Medicine, University of Pennsylvania, Philadelphia, Pennsylvania 19014; and Department of Obstetrics and Gynecology, Pennsylvania Hospital, Philadelphia, Pennsylvania 19107.   Alfred M. Bongiovanni • Departments of Pediatrics and Pediatrics in Obstetrics and Gynecology, School of Medicine, University of Pennsylvania, Philadelphia, Pennsylvania 19014; Department of Perinatal Endocrinology, Pennsylvania Hospital, Philadelphia, Pennsylvania 19107; and Department of Pediatrics, Thomas Jefferson University, Philadelphia, Pennsylvania 19107.

primordial germ cells. Germ cells first become recognizable in the 4-week embryo, when they may be identified embedded in the dorsal yolk sac epithelium. By the fifth week of embryonic life, they begin their migration to the ultimate site of the gonadal primordia. En route to this location, the genital ridge, the primordial germ cells undergo mitotic multiplication and continue to do so even after their arrival at the primitive gonad. The multiplication of these cells by mitotic division yields oogonia, which number approximately 6 or 7 million at maximum. During the course of their migration to the genital ridge, the germ cells contain no morphologically distinguishable features to characterize them as either male or female.[1] The coelomic epithelium possesses a thickened area located on the medial aspect of the mesonephros. The mesonephros protrudes into the coelomic cavity, carrying with it a supporting mesentery. The entire structure at this point in embryonic life, mesonephros and coelomic investment, is referred to as the urogenital ridge. The germ cells migrate into this primordial gonad via the mesentery of the gut. As the germ cells enter the genital ridge, they carry along with them some of the cells from the coelomic epithelium. The undifferentiated gonad thus consists of proliferating coelomic epithelium, a mesenchymal mass of cells located on the urogenital ridge which contains mesonephric components and recently arrived, still multiplying, primordial germ cells.

The ultimate development of the female duct system necessitates the *absence* of a factor responsible for disappearance of the embryonic mullerian ducts. In the embryonic male two distinct processes are responsible for (1) regression of the paired mullerian or paramesonephric ducts, and (2) differentiation of the paired mesonephric ductal system (Wolffian duct and mesonephric tubules). The two components are (1) mullerian regression factor and (2) testosterone. In the male, the mesonephric ducts develop into the epididymis, ductus deferens, and seminal vesicles. The singular role of the fetal testis in the process of male sexual differentiation was elegantly demonstrated by Jost,[2] who clearly proved that the fetal testis is necessary for three separate embryonic processes: 1) regression of the mullerian ducts, 2) maintenance of the Wolffian ducts, and 3) masculinization of the external genitalia. At the present time, two distinct factors are recognized as necessary for these three processes: one factor is required to achieve masculinization and Wolffian duct development; the second is essential to bring about mullerian duct regression. The factor responsible for regression of the mullerian duct, referred to as *mullerian regression factor* (MRF), is produced by the Sertoli cells in the fetal testicular seminiferous tubules. The second factor, the steroid hormone testosterone, originates from the Leydig cells of the male gonad. The presence of a surface antigen, present on all cells (H-Y antigen), has been associated with testicular gonadal development.[3] This antigen can be determined using immunologic techniques. Recently, the matter of the H-Y antigen has become somewhat confused because of several reports of its presence in phenotypic females with gonadal dysgenesis. These subjects have had the karyotype isochromosome of the long arm of the X.[4,5] Findings such as these suggest that the gene for the expression of the H-Y antigen may be located on an autosome and that it is in some way influenced by information on the X and the Y chromosomes. For example, in normal circumstances, it may be that X has a repressor and that Y has a "derepressor." At the present time, this matter is not entirely clear. In the majority of instances, however, it appears that H-Y antigen is associated with the presence of a Y chromosome. Human fetal testicular tissue accumulates testosterone in a pattern which suggests that the testes produce higher levels of androgen during that embryonic interval at which the male genitalia undergo differentiation.[6] Evidence for differentiation of the functions of these two distinct substances during embryonic development (MRF and testosterone) has been provided through the administration of an anti-androgen, cyproterone acetate, to pregnant rabbits and the subsequent demonstration that particular male characteristics are abolished in the male fetus exposed to this drug, namely internal genitalia derived from the Wolffian duct system and male external genitalia. Irradiation of human fetal testicular tissue tends to destroy germ cells while preserving Sertoli cells. Using this experimental approach, MRF is retained presumably because Sertoli cells are radioresistant and therefore able to persist. Thus, the factors that are responsible for male differentiation from the indifferent state are fairly clear. Considerably less information is available regarding the stimuli responsible for female sex differentiation. Information currently available would suggest that organization of the ovary and the mullerian duct system requires the presence of two X chromosomes, the lack of a Y chromosome, and the absence of MRF. With this background, it has been inferred that the primitive gonad has an intrinsic propensity to develop into an ovary, given the presence of germ cells and the lack of factors responsible for testicular differentiation. Recently, intriguing preliminary observations have been described based upon studies in the oophorectomized female monkey fetus that indicate that mullerian duct cilia are dependent for their development upon estradiol produced by the fetal ovary (R. M. Brenner, personal communication).[7]

2.1. Folliculogenesis

Organization of the gonad into an ovary has been said to begin with the formation of the first follicles. However, the ovary may be recognizable even earlier, at the time when germ cells enter meiotic prophase. Germ cells are distinguishable from other cells in the primordial gonad by virtue of their greater size, their round shape, nuclei containing a coarse chromatin network, and large acidophilic nucleoli. The female germ cells behave differently than those of the male. The female germ cells appear to be distributed throughout cords of cells that comprise the gonadal cortex. In contrast, the male germ cells tend to be located more centrally within the gonad, in the medullary region. By the 20th week of embryonic life, oogonia undergo transformation into oocytes. At this point in embryonic development, the possibility of any further increase in the germ cell complement is absolutely eliminated. Primordial follicles begin to appear shortly thereafter. These earliest follicles consist of an oocyte suspended in the prophase of the first meiotic division and surrounded by a single layer of flattened, squamous-like cells, separated from the surrounding stroma by a basement membrane. Transformation of these primordial follicles into primary follicles can be recognized by the acquisition of investing follicular cells, which are characterized by their cuboidal shape. Further proliferation of these cuboidal cells gives rise to a multilayered follicle. These cuboidal cells, which ultimately comprise the granulosa cell investment of the follicle, are readily distinguishable from the elongated and epithelioid peripheral cells, which ultimately become the theca interna.

Histologically, the follicles in fetal, postnatal, prepubertal, and postpubertal human ovaries are indistinguishable.[8] However, histochemical examination using specific stains demonstrates alkaline phosphatase activity in fetal follicles, which declines during early childhood. At puberty, the intensity of this alkaline phosphatase reaction once again increases. A similar pattern exists for glycogen content.[6] These observations suggest that follicular cell enzymatic activity is related to gonadotropic stimulation and resultant steroid biosynthesis, processes which are not in effect immediately after birth, but which are initiated at the time of puberty.

One of the most remarkable features of the oocytes once established within the ovary is that they remain suspended in prophase of the first meiotic division for many years. Resumption of the process of meiosis is ultimately prompted by the action of gonadotropin on the follicle.

Another unique feature of ovarian follicular development, also dependent upon gonadotropic activity, is the development of a well-defined cavity within the follicle, referred to as an *antrum*. When the antrum is recognizable, the follicle is said to have attained competence, having reached the graafian stage. At this point the cells peripheral to the granulosa layer have differentiated into two distinct layers: theca interna and theca externa. The outer layer, or externa, is relatively thin and its cells resemble connective tissue, blending into the surrounding stroma. The inner layer or theca interna is richly supplied by a capillary bed which does not penetrate the granulosa cell layer. In most instances throughout prepubertal and reproductive life, follicles undergo regression even before an antrum has begun to develop.[9]

2.2. Follicle Atresia

Follicle regression or atresia is an important process that in essence represents a natural limitation of reproductive capacity in the primate.

Follicular atresia occurs during fetal life, infancy, prepuberty, and following puberty, as well as during each ovulatory cycle, pregnancy, and throughout the use of ovulatory suppressants (birth control pills). During fetal life, follicular maturation occurs primarily among follicles in the deepest portions of the ovarian cortex, located near the medulla. At birth in the term female fetus, ovarian proportions measure approximately 10 × 2–4 mm. At this time, most of the germ cells have become incorporated into primordial follicles (Potter), although all stages of follicle development may be observed in the newborn, as well as various stages of follicle atresia. Although follicular atresia continues until the last oocyte disappears from the ovaries at the time of menopause, a decline in oocyte complement occurs rapidly during fetal life. Whereas the maximum number of germ cells at one time totaled 6 to 7 million in fetal life, at birth the ovaries contain 2 to 4 million oocytes. By menarche, only 400,000 oocytes are present in both ovaries. From birth and thereafter, the number of oocytes progressively diminishes. Each oocyte remains quiescent for many years; however, once it has been activated, by gonadotropin, the oocyte must either be released from the ovary through the ovulation process, or, by necessity, undergoes regression and ultimate demise. The long-term suspension of meiotic development in the ovary is achieved through inhibitory substances contained within the follicular antral fluid. When follicular development begins, this inhibitory process is either released or overcome, and the meiotic process resumes.

3. ENDOCRINOLOGY OF THE FEMALE FETUS

The hormonal environment present during embryonic and fetal life is partially responsible for male/female differentiation. This hormonal milieu is unique, differing significantly in the female from that which prevails after birth and throughout adulthood. The maturation of oocytes and follicles within the ovaries is dependent upon gonadotropic stimulation. During infancy and prepubertal childhood, pituitary gonadotropin (follicle-stimulating hormone, FSH, and luteinizing hormone, LH) secretion is minimal, and ovarian function appears to proceed at a relatively low level. In contrast, during fetal life three potential sources of gonadotropins can be recognized: the fetal pituitary gland, the maternal pituitary gland, and the placenta. Gonadotropins from each of these sources may stimulate fetal ovarian activity.

Both FSH and LH make their appearance in the fetal pituitary gland by the tenth gestational week, and by 12 weeks FSH can be detected in fetal serum.[10] By midgestation, fetal serum LH and FSH have attained peak levels and thereafter decrease. At term, the levels of LH and FSH present in umbilical venous blood are low. Gonadotropin levels in the pituitary gland itself are of particular interest. Fetal pituitary gland FSH content and fetal serum FSH levels are greater during midgestation in females than in males. This difference in pituitary and serum FSH levels between male and female fetuses probably reflects the high concentrations of fetal testosterone in the male fetus attained during midgestation. As a consequence, during this period of elevated serum testosterone levels in the male fetus, the synthesis, storage, and secretion of FSH by the pituitary gland are reduced. This inhibitory effect of fetal testosterone on FSH production probably represents one of the earliest examples of the negative feedback mechanism between gonadal and pituitary secretions. Gonadotropin-releasing hormone (GnRH) has also been detected in the human fetal hypothalamus early in pregnancy. Fetal pituitary LH and FSH synthesis and secretion during fetal life are thus likely to be under regulation of GnRH produced by the hypothalamus. This concept is supported by the observation that fetal pituitary gonadotropic cells in culture respond to GnRH. Grumbach and Kaplan[10] have postulated that the hypothalamic–pituitary–gonadotropic–sex steroid negative feedback mechanism exhibits increasing sensitivity throughout pregnancy. With the establishment of such a regulatory system, FSH and LH secretion persist unrestrained in midgestation. As responsiveness of the pituitary to inhibitory CNS influences and as the degree of negative feedback by sex steroids increase, FSH and LH levels decline. Thus, with progression of fetal development, the negative feedback mechanism between gonad and hypothalamus–pituitary matures, the hypothalamus secretes less GnRH, and FSH and LH secretion consequently decline. This inhibition of hypothalamic GnRH elaboration reflects an increasing sensitivity of the hypothalamus to peripheral influence and also represents the very first indication of a "gonadostat" mechanism, which may be particularly instrumental in establishing the precise time at which puberty begins.

During human pregnancy, the maternal pituitary gland normally undergoes enlargement. Throughout the course of pregnancy, pituitary microadenomata also tend to exhibit considerable growth. However, the fact that successful pregnancy, labor, and delivery of healthy offspring can occur in the hypophysectomized female suggests that the maternal pituitary gland is not essential during pregnancy. Concomitant with increasing secretion of hCG by the placenta during pregnancy, maternal serum concentrations of FSH decline.[11] Levels of maternal FSH during pregnancy are clearly lower than those observed throughout the menstrual cycle.[12] The maternal concentration of FSH in pregnancy contrasts with FSH levels in the fetus. Even the male fetus demonstrates higher serum FSH levels than those found in maternal serum. During pregnancy LH levels in maternal serum are likewise reduced.[13–15] Maternal LH responsiveness to GnRH in pregnancy decreases progressively, and FSH fails to respond to intravenous GnRH.

The trophoblast is the third source of gonadotropins to which the fetus is exposed. Human chorionic gonadotropin possesses both LH- and FSH-like activity. The time of maximal placental hCG production throughout pregnancy (10 to 12 weeks) coincides with that period in the male fetal development during which testicular testosterone production reaches its maximum. The fetal testis binds hCG and also produces testosterone in response to hCG.[16] Chorionic gonadotropin is probably influential in the regulation of fetal testicular steroid production. Testosterone production depends upon gonadotropic stimulation, and fetal testosterone plays a major role in male differentiation. Nonetheless, the fetal testis is able to produce testosterone under conditions in which fetal pituitary function is diminished. For example, male anencephalic fetuses demonstrate a marked reduction in Leydig cell complement and decreased weight of testosterone-responsive accessory tissues.[17] Although the human fetal pituitary gonadotropins are not essential for sex differentiation, it appears that they do play a role in the complete differentiation and development of the testes and ovaries. Thus, the male anencephalic infant may have hypoplastic external genitalia, underdeveloped epididymis, and a decreased number of Leydig cells. [10,18,19] The fact that fetal testicular testosterone production occurs to any degree in the absence of fetal pituitary gonadotropins provides compelling evidence that placental hCG possesses testicular-stimulating properties.

Ovarian morphology in the female anencephalic fetus is reported to be indistinguishable from that of the normal fetus at comparable gestational age; however, ovarian weight has been reported to be significantly reduced in comparison with normals.[20] These two controversial observations serve to indicate the need for more information regarding gonadotropin regulation of fetal ovarian development. The recognized dependence of follicular development and estrogen biosynthesis upon FSH activity suggests that the elevated levels of FSH found in the fetal female serum during midgestation may be responsible for presence of mature follicles within the ovaries. Reduced numbers of graafian follicles found in the ovaries of the anencephalic fetus also provides evidence that FSH stimulation is required to achieve follicular maturation beyond the primary stage.[17] Females born with olfactogenital dysplasia (Kallman syndrome) exhibit hypogonadotropic hypogonadism and primary amenorrhea. Individuals so afflicted are inclined to have a eununchoid habitus with tall stature, infantile external genitalia, and hypoplastic uteri. The ovaries of such women characteristically lack graafian follicles. Although primordial follicles can be found with advancement to the primary follicle stage, this represents the maximum degree of follicular maturation achieved in this condition.[21] A recent case report has verified that, despite lack of ovarian stimulation by gonadotropins during fetal life in Kallman syndrome, pituitary gonadotropin secretion can be provoked by administration of GnRH[22] and ovarian tissue is capable of responding in turn to the endogenous gonadotropins. This observation implies that pubescence is initiated by maturation of neuroendocrine centers located above the level of the pituitary gland. Furthermore, ovaries containing follicles that have never developed further than the primary stage are nonetheless capable of gonadotropin stimulation, maturation, ovulation, and release of ova capable of fertilization.

Limited data are available to indicate that the fetal ovary produces estradiol *in utero*. Reyes et al.[23] observed higher estradiol levels in female fetal serum as compared with male fetuses between the 10th and 24th weeks of gestation, while Robinson et al.[24] have reported higher estradiol levels in amniotic fluid of the female than of the male.[24] Ciliated epithelium is normally found lining the Mullerian ducts in the fetal monkey. Ciliation of these epithelial cells in the adult is known to be dependent upon estradiol.[4] Oophorectomy performed on the monkey fetus *in utero* results in failure of ciliation of the mullerian ducts (R. M. Brenner, personal communication).[7] The endocrine activity of the fetal ovary requires more investigation before its function can be fully elucidated. Evidence that secondary sexual structures (e.g., vagina, cervix, breast) in the female fetus and newborn have already been stimulated indicates the sensitivity of receptive cells in these tissues to estrogens, which may originate in the fetus, mother, and/or placenta.

3.1. Endocrinology of the Newborn Female

During the first few months of life in healthy infants, the pituitary–gonadal system demonstrates evidence of transient activity. Full-term males and females both display elevations in serum levels of FSH and LH.[25] These gonadotropin levels in turn provoke the stimulation of testicular production of androgen in males and an increase in ovarian steroid elaboration in females. The pattern differs in male and female infants. The elevation of FSH is higher and more sustained in female than in male infants. Low levels of serum FSH and LH characteristic of childhood are attained in boys by 4 months of age, while in females such low levels of gonadotropins are not achieved until the age of 4 years. According to Winter and Faiman,[26] the declining levels of gonadotropins are accomplished through the inhibitory feedback influence of infantile gonadal steroid production. This hypothesis is based upon the observation that in a group of agonadal subjects serum FSH concentrations are elevated above those anticipated at any given age from 2 months to 22 years. Serum LH levels were also elevated in most, but not all, of these agonadal subjects. A further increment takes place in serum gonadotropin concentrations after the normal age of puberty in agonadal individuals. These changes in the sensitivity of the hypothalamus during normal puberty may proceed even in the absence of a functional gonad. Ovarian sex steroid secretion in female infants occurs after birth in response to elevated serum FSH and LH levels. This observation naturally leads to the question of whether the temporary activation of pituitary–gonadal stimulation is (1) a function of the withdrawal of the negative feedback effects of placental steroids immediately after birth or (2) the result of maturation of fetal pituitary endocrine capacity itself. A comparison of pituitary–gonadal function in full-term and premature girls demonstrates serum FSH levels 10–20 times higher in premature than in full-term infants; similarly LH levels are 3–4 times higher in premature than in term girls.[27] The last weeks of gestation must be significant in the functional maturation of the fetal hypothalamic–pituitary–gonadal system, since premature interruption of the process by preterm birth gives rise to enhanced pituitary gonadotropin production in girls. A progressive sensitization of the hypothalamic–pituitary regulatory system, or gonadostat, to sex steroids must occur during the last few weeks of pregnancy. The prematurely born infant possesses low sensitivity of the gonadostat and fails to respond to her low circulating estrogen levels with suppression of gonadotropins. Thus, values of serum estradiol in the female infant at 6 months are still above the later childhood range and do not fall completely until 4 years of age.[25]

3.2. Ovarian Morphology in the Female Infant

The ovary grows from the time of birth until menarche. Ovarian growth depends primarily on an increase of stromal tissue and blood vessels. The growth in reticular fibers is accompanied by a marked decrease in the number of oocytes.[8] The greatest decline in oocyte complement occurs during the first decade of life.[28] Most of these oocytes die during the earliest phases of follicular development before the stage of antrum formation. By the age of 11 or 12, the ovary is composed of a central zone of connective tissue containing many blood vessels and a bipartite peripheral zone. The superficial zone of the ovary is hyalinized and pale, while the deeper zone is darker and more cellular. Primordial follicles can be found evenly distributed at or near the junction of these two cell layers. Follicles at various stages of development surround the medulla, encroaching on both cortex and medulla during growth. The surface epithelium or germinal epithelium is composed of a single layer of inactive, functionless cells.[9] The major difference between the morphologic appearance of the ovary during late childhood and throughout postpubertal development is that antral stage follicles and corpora lutea are lacking in the former.

4. ADOLESCENT GROWTH AND MATURATION

There is an increase in the rate of growth (both height and weight) with the onset of puberty.[29] This phenomenon is highly characteristic and is of importance to the clinician, among other findings in evaluating the girl of pubertal age. There are three phases. The first is one of minimum growth velocity, sometimes call the "age of take-off," which represents more or less the same velocity as in the 3 or 4 years prior to the onset of puberty. This is followed by a period of the most rapid growth or the "peak height velocity." Finally, there is a period of decreased velocity with the cessation of growth and fusion in the epiphyses. Girls reach the peak height velocity before the menarche so that any further growth in the postmenarchal girl is limited. The first stage begins at an average age of 11 years with the peak velocity at around 12 years and completion shortly before the 14th year. The matter of the increase in body weight or "fatness" is emphasized by Frisch and her co-workers.[30,31] It is felt that the onset of menstruation in girls depends upon the attainment of this ideal weight, which in some ways affects the hypothalamus thus disinhibiting its secretion of GnRH. This hypothesis has some merit. Thus, it is known that the age of menarche occurs earlier in most societies with better nutrition and greater weight at a given age. This argument is also reinforced in the cessation of menstruation in anorexia nervosa, although such cessation is often known to occur before the loss of weight.

The secondary sexual development is a clinical index to the status of a pubertal girl and if well categorized often eliminates the need for the measurement of hormones. A useful compendium designed from Marshall and Tanner[32] is shown in Table I. The development of the breasts and the external genitalia in the girl are closely related to ovarian secretions. However, other factors are also at play particularly with regard to the development of the breasts.

Stage I reveals no development of the secondary sexual characteristics and generally lasts throughout the first decade of life. In American girls somewhere between $10\frac{1}{2}$ and 11 years, there is very early budding

TABLE I. Stages of Pubertal Development in Girls[a]

Stage		Chronologic age (years)
I	No signs of sexual development	Under 11
II	Early budding of breasts confined to periareolar region with a few tubercles; few pubic and/or axillary hairs; early prominence of labia minora and majora; initiation of increased growth velocity	10.5 ± 2
III	Increased fullness of breasts with projection of areolae and nipples, many tubercles; small amount of pubic and axillary hair; moderate enlargement of labia minora and majora; dulling vaginal mucosa; maximal height velocity; growth of clitoris; vagina enlarges, pH becomes acid (4–5); menarche in 30%	11.5 ± 2
IV	Breasts and external genitalia well developed; moderate to abundant pubic hair; onset of menarche	12.5 ± 3
V	Adult sexual development; ovulation; deceleration of growth	14 ± 3

[a] Practical approximations based on several sources for American girls.

of the breasts which is noted as a slight fullness behind the nipples and confined to this region with perhaps a few tubercles. At this time there may appear a very few downy hairs primarily along the labia majora. And there is a slight prominence of both the labia majora and minora, which was not evident earlier. In Stage III, there is a further enlargement of breasts and areolae but there is no separation of the contours. In Stage IV, the breasts enlarge further, and there is now a projection of the nipples so that a secondary mound is formed. The pubic hair spreads farther over the pubes but encompasses a smaller area than in the adult. In Stage V, the breasts now resemble those of the mature woman, and the areolae recess again into the general contour of the breast. Menarche usually occurs in Stage IV, although it may be somewhat earlier. The changes in the mammae and pubic hair are outlined in Chapter 10, Figs. 1 and 2, and Table I.

Osseous maturation proceeds from fetal life continuously through childhood into the age of complete maturation. It is measured by employing radiographs of the wrist (sometimes together with the knee and elbow) and assigning an age according to available standards.[33,34] The osseous maturation is a more accurate guide to the developmental status of the growing girl than is the chronologic age. Thus in Table I, it would be more accurate to indicate that the described changes of adolescence will conform better to the osseous than the chronologic age. This technique is of great assistance in the evaluation of "early" or "late" maturation. The changes in the osseous maturation occurring around the time of puberty are in great measure related to the ovarian secretions and in small degrees to the adrenal androgens. Together with the current height and chronologic age, bone age is employed for the estimation of this final adult height.[35,36]

There is also a maturation of the adrenal cortex at the time of puberty, which probably explains the development of the sexual hair and the enlargement of the clitoris. It has been known for some years that the level of the adrenal androgens in the blood rise remarkably at this time and quantitatively represent the most dramatic alteration of the steroid hormones.[37,38] This participation of the adrenal normally precedes the increase of gonadotropins and ovarian hormones by 2 years. It may occur earlier than this. The major steroids secreted at this time by the adrenal cortex are dehydroepiandrosterone (DHEA) and its sulfate (DHEAS). Androstenedione also increases at this time and 50% of this steroid arises from the adrenal, the remainder from the ovary. The latter contributes to circulatory testosterone and estrone by peripheral conversion.

The exact mechanism for the maturation of the adrenal cortex at this time remains unclear.[39] A popular hypothesis has been the secretion of a distinct pituitary adrenal androgen-stimulating hormone. There is frequently seen in clinical practice the precocious appearance of sexual hair which is not associated with the other features of secondary sexual development. This has been termed *premature adrenarche*. It does not appear that these adrenal androgens are essential for the adolescent growth spurt, whereas the ovarian steroid in conjunction with growth hormone are important. The normal values according to age and stage of development of the adrenal secretions are tabulated in Chapter 10.

5. HORMONAL CHANGES AT PUBERTY

5.1. CNS

The hypothalamus, pituitary, and ovaries have a unique interdependent relationship. As a result, it has been difficult to identify which of these structures is the first to attain maturity. Recent studies, however, point to the CNS as the limiting element in determining the onset of puberty. Puberty in the female may be defined as that stage in development at which the ovaries secrete sex hormones, primarily estradiol, in quantities sufficient to stimulate growth of genital structures and produce secondary sexual characteristics. Ovarian estrogen secretion is responsible for other important physiologic changes, including acceleration of linear growth and skeletal maturation.

During pubescence, many changes in hormonal secretion occur simultaneously, including increased elaboration of pituitary gonadtropins, ovarian steroids, and adrenocortical steroids. Adrenal androgens (DHEA and DHEAS) progressively increase early in puberty. This finding suggests the possibility that adrenal androgens may initiate CNS activation during puberty.[40] Agonadal females (e.g., subjects with Turner syndrome) achieve elevated concentrations of FSH and LH well before the anticipated time of puberty. Thus, it is likely that in normal circumstances ovarian estrogens are responsible for suppressing pituitary gonadotropin secretion even before puberty begins. The fact that an additional rise in gonadotropin secretion occurs in the agonadal female at the time when puberty might be anticipated suggests further functional changes in the feedback system that occur at the time of puberty.

The pineal body has also been implicated as a factor in the onset of puberty. Pineal secretions inhibit release of pituitary gonadotropins prior to maturity. Most currently held concepts regarding pineal function stem from animal experiments. Ablation of the pineal gland of the cock at an early age leads to increased weight of the testes and comb with concomitant increase in gonadotropin potency.[41] In contrast, administration of pineal extracts results in decreased gonadal weight. Wurtman and co-workers[42] demonstrated that melatonin, a substance synthesized by the pineal gland, lowers the incidence of estrus in the rat and slows ovarian growth. The presence of light decreases pineal production of melatonin. If this action is dependent upon light receptors in the retina, one might anticipate retarded sexual development in blind children. The reverse sequence has been observed, however, and girls blind from birth experience menarche earlier than those with acquired blindness.[43] The role of the pineal gland in puberty in the human is still not fully understood.

Environmental factors other than light also influence the timing of puberty. Diverse influences, such as emotional stress, body weight, and body fat, undoubtedly exert their effects on centers in the brain which ultimately regulate hypothalamic activity. A relationship has been demonstrated between height and body weight and menarche[30] Although early- and late-maturing girls experience menarche at the same mean weight, late maturers tend to be taller at menarche. This observation leads to the proposition that attainment of a critical body weight is essential for adolescent events. Gonadal hormones in very small amounts restrain the secretion of pituitary gonadotropins during early childhood. In order to permit greater interplay between the pituitary and ovaries, increasing amounts of gonadotropic hormone are required to evoke ovarian stimulation. This increase may be accomplished via enhanced pituitary gonadotropin release either through the action of stimulatory phenomena, e.g., GnRH, or via lessened sensitivity of the hypothalamic–pituitary system to steroid inhibition. As maturation progresses, increased amounts of gonadotropins are secreted, more ovarian steroids are produced, and the characteristic changes associated with puberty proceed.

5.2. Changes in Pituitary Function at Puberty

Lee et al.[44] showed a 3.5-fold rise in LH and 1.7-fold rise in FSH when they compared prepubertal children and postpubertal adolescents. The relatively high levels of serum LH and FSH in girls between ages 11 and 14 as compared to values in boys correlates with the earlier chronology of pubertal changes in girls. A distinction must be made in evaluating gonadotropin levels between immunoassayable and bioassayable gonadotropins. The preceding statements refer to immunoreactive FSH and LH. Recently it has been demonstrated that whereas LH immunoreactivity in blood increases about 3-fold from prepuberty to late puberty, the bioactivity increases 8.2-fold. The mechanism for this qualitative change is unknown. Although studies have been conducted in boys, the same may be true of girls. The relatively greater increase in the LH bioactivity during puberty, at least in males, would explain the discrepancy between low basal immunoreactive LH and normal testosterone concentrations found in boys with isolated growth hormone deficiency.[47] Although immunoassayable LH is detectable in the serum of prepubertal girls, the levels of bioassayable LH are not usually detectable. As in boys, the girls will show a much greater rise in the bioassayable hormone between puberty and adulthood than in the immunoassayable hormone. The reasons for these discrepancies are not entirely clear at present. However, these observations suggest that the radioimmunoassay may not accurately reflect the true secretion of the gonadotropins during puberty and adolescence.[46,47] Pituitary secretion of LH and FSH is regulated by GnRH, and in the prepubertal female, low serum concentrations of ovarian steroids restrain GnRH and LH and FSH secretion at low levels, a reflection of the high degree of sensitivity inherent in the hypothalamic receptors. These receptors gradually and progressively become less sensitive to the negative feedback action of the steroids. As a result, GnRH elaboration increases; in turn GnRH induced LH and FSH synthesis and secretion increase, stimulating ovarian steroidogenesis. This progressive change in sensitivity of the hypothalamus continues until the adult relationship between the ovaries and the hypothalamic–pituitary complex is attained. The term *gonadostat* has been applied to this delicate interaction.[45]

Gonadotropins are elaborated in pulsatile fashion. Prior to puberty, minor fluctuations in LH and FSH levels can be recognized throughout the course of the day. The earliest endocrinologic evidence of puberty may be observed in the surges of LH secretion occurring during sleep.[48] Late in puberty, marked oscillations in serum gonadotropins occur during both day and night. Ultimately an adult pattern of gonadotropin pulsations is attained. Gonadotropin secretion in anorectic patients resembles that observed in prepubertal girls. In anorexia nervosa, basal levels of LH are low and nocturnal LH secretion is conspicuously absent. As body weight is restored to normal, serum LH and FSH levels gradually increase. In this process, patterns of LH and FSH sequentially pass through the stages characteristically found during puberty until mature adult patterns are once again achieved. Marshall and Kelch[49] studied patients with anorexia nervosa to determine whether GnRH can induce the hormonal changes found in normal puberty. The responses observed simulated those associated with normal puberty in girls. We can interpret these findings to indicate the dependence of changes in gonadotropin secretion during normal puberty upon the effects of GnRH. In normal girls, the concentration of LH in serum increases with advancing skeletal maturation and so also does the response to the administration of GnRH. In contrast, the FSH responses to the GnRH show a linear negative correlation with advancing bone age.[50] Estradiol levels in blood have been found to rise slowly and steadily with advancing sexual maturation, correlating appropriately with bone age, chronologic age, and clinical parameters of pubertal development.[51]

The secretion of LH and FSH from the anterior lobe of the pituitary is dependent upon the hypothalamic secretion of GnRH. This substance, a simple decapeptide, originates in the medial basal hypothalamus (MBH) and reaches the pituitary gland by way of the hypothalamic–hypophyseal portal system of vessels. As is the case with FSH and LH, hypothalamic release of GnRH is pulsatile. The oscillations, however, are not necessarily in phase with those of the pituitary gonadotropins. As demonstrated by Nakai et al.,[52] if the mass of the CNS attached to the pituitary gland is reduced, pituitary function persists undisrupted, provided that the MBH connection to the pituitary remains intact. The arcuate nucleus serves as the source of GnRH. When lesions are created in this area, FSH and LH secretion falls abruptly and both negative and positive feedback mechanisms are abolished. An "artificial hypothalamus" in the rhesus monkey has been created by destruction of the arcuate nucleus and infusion of GnRH directly into the pituitary gland. Using this model, an abrupt decline in LH and FSH follows creation of the lesion in the arcuate nucleus. Infusion of GnRH into this system produces a peak in FSH and LH followed by a decrease which persists during continuous GnRH infusion. If GnRH is infused in an intermittent or pulsatile fashion, both FSH and LH will be secreted in a pulsatile pattern. This pulsatile scheme of GnRH infusion also restores the ability of estradiol to provoke appropriate LH and FSH responses, either positive or negative. Such experiments conducted in this unique model imply that both positive and negative feedback actions of estradiol are exhibited at the level of the pituitary gland, while the role of the GnRH is permissive. The need for intermittent exposure of the pituitary gland to GnRH for appropriate gonadotropin secretion reflects the concept of "down regulation." In this mechanism, a small amount of agonist (GnRH) is required at specific intervals to achieve its effect; exposure to the agonist for protracted intervals overwhelms the receptors and leads to a refractory state. Taken one step further, Knobil et al.[53] produced ovulatory cycles, as evidenced by FSH and LH patterns and plasma progresterone concentration resembling those characteristic of normal spontaneous menstrual cycles, in rhesus monkeys with hypothalamic lesions, treated with GnRH administered in a pulsatile fashion. Despite a constant rate and schedule of GnRH infusion, the pattern of pituitary gonadotropin secretion paralleled that of the normal menstrual cycle, suggesting that gonadotropin secretion is not governed by alterations in patterns of hypothalamic GnRH secretion but rather by ovarian estradiol production, which ultimately acts directly on the pituitary gland. Subsequently, a similar model was established in prepubertal female rhesus monkeys. The infusion of GnRH for 6 min per hour initiated ovulatory cycles in all six infantile animals studied.[54] Initially gonadotropins became detectable in the serum within several days of initiation of GnRH pulsatile infusions. Plasma estradiol subsequently increased to levels characteristic of follicular phase levels. The initial peaks in estradiol (in excess of 200 pg/ml) failed to induce preovulatory surges of gonadotropins, although subsequent peaks in estradiol provoked surges in LH which led to the appearance of progesterone in the plasma. These observations support the conclusion that both pituitary and ovarian competence have been established well before the neuroendocrine factors controlling pulsatile GnRH release from the hypothalamus have matured sufficiently to direct the orderly sequence required for adult patterns of ovarian function.

5.3. Menarche

The earliest menstrual cycles during attainment of puberty are anovulatory and simply represent exposure of the uterine endometrium to ovarian estradiol. During the first postmenarchal year the responses of the serum gonadotropins and the sex steroids to the administration of GnRH do not yet reach adult levels. Between the third and the fifth postmenarchal years, there is a progressive increase in the response and they are indeed greater than in normal adult females. Nonetheless, measurements of serum progesterone indicate that there is still a low percentage of ovulatory cycles. The levels of estradiol, however, are within the range of normal adults, and this causes an exaggerated response to the GnRH stimulation. When these cycles, however, become ovulatory with levels of progesterone, which are greater than 10 ng/ml, a lower response to GnRH stimulation, which is more in keeping with adult behavior, is observed. Therefore, it would appear that high concentrations of progesterone may exert a negative effect on the response to GnRH,[55] In the early stages of puberty in the female, as mentioned elsewhere, there is a tonic circadian secretion of gonadotropins, which eventually is converted to the cyclic adult pattern. Although the responses to GnRH are shown to increase gradually with the progress of maturation, as detailed in Chapter 10, some other noteworthy peculiarities have recently been reported. The response to GnRH stimulation can often not be correlated with either the chronologic or the bone age. However, there would appear to be a good correlation with the maturation of the breasts. In particular, there is a remarkable response of the LH levels in serum with such stimulation and, as noted in other studies, this responsiveness diminishes after the onset of the menarche. The work of Illig et al.[56] suggests a role for estradiol that is similar to the observations of others. Estradiol influences breast development, but it also acts directly on the pituitary. This action may serve to sensitize the gland to stimulation by releasing hormone. Then with the onset of regular ovulation, it is possible that progesterone diminshes the sensitivity.[56] Once full maturation of the hypothalamic–pituitary–ovarian complex has been achieved, in the normal female, cycles become ovulatory. Following an estradiol peak, a surge in gonadotropins leads to follicular disruption, ovulation, conversion of the follicle to a corpus luteum, and progesterone production. In the absence of pregancy, corpus luteum function persists for approximately 2 weeks. With the waning of the corpus luteum, progesterone secretion declines and progesterone withdrawal flow occurs. Establishment of ovulatory cycles signifies sexual maturity in the female as indicated by achievement of reproductive competence.

REFERENCES

1. Baker TG: A quantitative and cytological study of germ cells in human ovaries. *Proc Roy Soc (Biol)* 158:417–433, 1963
2. Jost A: Recherches sur la differenciation sexuelle de l'embryon de lapin. 3. Role des gonades foetales dans la differenciation sexuelle somatique. *Arch Anat Microscop Morphol Exp* 36:271–273, 1947
3. Silvers WK, Wachtel SS: H-Y antigen: Behavior and function. *Science* 195:956–960, 1977
4. Wolf U, Fraccaro M, Mayerova A, Hecht T, et al: Turner syndrome patients are H-Y positive. *Hum Genet* 54:315–318, 1980
5. Meade KW, Wachtel SS, Davis JR, et al: H-Y antigen in XO X, Iso (X) mosaic Turner syndrome. *Obstet Gynecol* 57:594–599, 1981
6. Reyes FI, Winter, JSD, Faiman C: Studies on human sexual development. 1. Fetal gonadal and adrenal sex steroids. *J Clin Endocrinol Metab* 37:74–78, 1973
7. Brenner RM, Anderson RGW: Endocrine control of ciliogenesis in the primate oviduct, in Greep, RO, Astwood, EB (eds): *Handbook of Physiology*, Section 7, Vol. II, *Endocrinology* Baltimore, Williams & Wilkins 1973, pp 123–140
8. Pinkerton JHM, McKay DG, Adams EC, et al: Development of the human ovary: A story using histochemical techniques. *Obstet Gynecol* 18:152–181, 1961
9. Potter EL: The ovary in infancy and childhood. In Grady, HG, Smith, DE (eds): *The Ovary*, Chapter 2. Baltimore, Williams & Wilkins, 1963, p 11
10. Grumbach MM, Kaplan SA: Fetal pituitary hormones and the maturation of central nervous system regulation of anterior pituitary function, in Ulick, L (ed), *Modern Perinatal Medicine*, Chicago, Year Book Med. Publ., 1974, pp 247–271
11. Jaffe RB, Lee PA, Midgley AR, Jr: Serum gonadotropins before, at the inception of, and following human pregnancy. *J Clin Endocrinol Metab* 24:1281–1283, 1969
12. Takagi S, Yoshida T, Tsubata K: An investigation of the maternal fetal hormonal milieu with special emphasis on the maternal influence, in Notake, T, Suzuki, S (eds) *Biological and Clinical Aspects of the Fetus*. Tokyo, Igaku Shoim, 1977 p 134
13. Sowers J, Colantino M, Fayez J, et al: Pituitary response to LHRH in midtrimester pregnancy. *Obstet Gynecol* 52:685–688, 1978
14. Reyes FI, Winter JSD, Faiman C: Pituitary gonadotropin function during pregnancy: Serum FSH and LH levels before and after LHRH administration. *J Clin Endocrinol Metab* 42:590–592, 1976
15. Miyake A, Tanizawa O, Anono T, et al: Pituitary responses in LH secretion to HRH during pregnancy. *Obstet Gynecol* 49:549–551, 1977
16. Huhtaniemi IT, Korenbrot CC, Jaffe RB: HCG binding and stimulation of testosterone biosynthesis in human fetal testes. *J Clin Endocrinol Metab* 44:963–967, 1977
17. Blizzard RM, Albert AM: Hypopituitarism, hypoadrenalism and hypogonadism in the newborn infant. *J Pediat* 48:782–792, 1956
18. Bearn JG: Anencephaly and the development of the male genital tract. *Acta Paediatr Acad Sci Hung* 9:159–180, 1968
19. Zondek LH, Zondek T: The influence of complications of pregnancy and of some congenital malformations in the reproductive organs of the male foetus and neonate. *Int Symp Sexual Endocrinol Perinatal Period* 32:79, 1974

20. Ross GT: Gonadotropins and preantral follicular maturation in women. *Fertil Steril* 25:522–543, 1974
21. Sparkes RS, Simpson RW, Paulsen CA: Familial hypogonadotropic hypogonadism with anosmia. *Ann Int Med* 121:534–538, 1968
22. Soules MR, Hammond CR: Female Kallmann's syndrome: Evidence for a hypothalamic luteinizing hormone-releasing hormone deficiency. *Fertil Steril* 33:82–85, 1980
23. Reyes RI, Boroditsky RS, Winter JSD, et al: Studies on human sexual development. 2. Fetal maternal serum gonadotropin and sex steroid concentrations. *J Clin Endocrinol Metab* 38:612–617, 1974
24. Robinson JD, Judd HL, Young PE, Jones OW, et al: Amniotic fluid androgens and estrogens in midgestation. *J Clin Endocrinol Metab* 45:755–761, 1977
25. Winter JSD, Hughes IA, Reyes FI, et al: Pituitary-gonadal relations in infancy. 2. Patterns of serum gonadal steroid concentrations in man from birth to two years of age. *J Clin Endocrinol Metab* 42:679–686, 1976
26. Winter JSD, Faiman C: Serum gonadotropin concentrations in aponatal children and adults. *J Clin Endocrinol Metab* 35:561–564, 1972
27. Tapanainen J, Koivisto M, Vihko R, et al: Enhanced activity of the pituitary-gonadal axis in premature human infants. *J Clin Endocrinol Metab* 52:235–238, 1981
28. Valdes Dapena MA: The normal ovary of childhood. *Ann NY Acad Sci* 142:597–613, 1967
29. Tanner JM, Whitehouse RH, Marubini E, et al: The adolescent growth spurt of boys and girls of the Harpenden growth study. *Ann Human Biol* 3:109–126, 1976
30. Frisch RE, Revele R: Height and weight at menarche and a hypothesis of critical body weights and adolescent events. *Science* 169:397–399, 1970
31. Frisch RE: Fatness of girls from menarche to age 18 with a nomogram. *Human Biol* 48:353–359, 1976
32. Marshall WA, Tanner JM: Variations in pattern of pubertal changes in girls. *Arch Dis Child* 44:291–303, 1969
33. Greulich WW, Pyle SI: *Radiographic Atlas of Skeletal Development of the Hand and Wrist*, ed 2. Stanford, California: Stanford Univ. Press, 1959
34. Tanner JM, Whitehouse RH, Marshall WA, et al: *Assessment of Skeletal Maturity and Prediction of Adult Height*, New York, Academic Press, 1975
35. Bayley N, Pinneau SR: Tables for predicting adult height from skeletal age: Revised for use with the Greulich-Pyle standards. *J Pediatr* 40:423–441, 1952
36. Roche AF, Wainer H, Thissen D: The RWT method for the prediction of adult stature. *Pediatrics* 56:1027–1033, 1975
37. Savage DC, Forsyth CC, McCafferty E, et al: The excretion of individual adrenocortical steroids during normal childhood and adolescence. *Acta Endocrinol* 79:551–567, 1975
38. Reiter EO, Fuldauer VG, Root AW: Secretion of the adrenal androgen, dehydroepiandrosterone sulfate during normal infancy, childhood and adolescence, in sick infants and in children with endocrinologic abnormalities. *J Pediatr* 90:766–770, 1977
39. Grumbach MM, Richards HE, Conte FA, et al: Clinical disorders of adrenal function and puberty: An assessment of the role of the adrenal cortex in normal and abnormal puberty in man and evidence for an ACTH-like pituitary adrenal androgen stimulating hormone, in *Serono Symposium on The Endocrine Function of the Human Adrenal Cortex*. New York, Academic Press, 1977, p 115
40. Hooper BR, Yen SSC: Circulating concentrations of dehydroepiandrosterone and dehydroepiandrosterone sulfate during puberty. *J Clin Endocrinol Metab* 40:458–461, 1975
41. Shellabarger CJ: Observations of the pineal in the white leghorn capon and cockrel. *Poultry Sci* 32:189, 1967
42. Wurtman RJ, Axelrod J, Chu EW: Melatonin, a pineal substance: Effect on the rat ovary. *Science* 141:277–278, 1963
43. Zacharias L, Wurtman RJ: Blindness: Its relation to age of menarche. *Science* 144:1154–1155, 1964

44. Lee Pa, Midgley AR, Jr, Jaffe RB: Regulation of human gonadotropins. VI. Serum follicle stimulating and luteinizing hormone determinations in children. *J Clin Endocrinol Metab* 31:248–253, 1970

45. Grumbach MM, Roth JE, Kaplan SL, et al: Hypothalamic-pituitary regulation of puberty evidence and concept derived from clinical research, in Grumbach, MM, Grave, GD, Mayer, PE (eds): *Control of the Onset of Puberty*, New York: Wiley, 1974, pp 115–166

46. Reiter OE, Beitins IZ, Ostrea T, et al: Bioassayable luteinizing hormone during childhood and adolescence and in patients with delayed pubertal development. *J Clin Endocrinol Metab* 54:155–161, 1982

47. Lucky AW, Rich BH, Rosenfield RL, et al: LH bioactivity increases more than immunoreactivity during puberty. *J Pediatr* 97:205, 1980

48. Weitzman ED, Boyar RM, Kapen S, et al: The relationship of sleep and sleep stages to neuroendocrine secretion and biological rhythms in man. *Rec Progr Horm Res* 31:399–446, 1975

49. Marshall JC, Kelch RP: Low dose pulsatile gonadotropin-releasing hormone in anorexia nervosa: A model of human pubertal development. *J Clin Endocrinol Metab* 49:712–718, 1979

50. Sauder SE, Corly KP, Hopwood NJ, et al: Subnormal gonadotropin responses to gonadotropin-releasing hormone persist into puberty in children with isolated growth hormone deficiency. *J Clin Endocrinol Metab* 53:1186–1192, 1981

51. Jenner MR, Kelch RP, Kaplan SL, et al: Hormonal changes in puberty. IV. Plasma estradiol, LH, and FSH in pre-pubertal children, pubertal females and in precocious puberty, premature thelarche, hypogonadism and in a child with a feminizing ovarian tumor. *J Clin Endocrinol Metab* 34:521–530, 1972

52. Nakai Y, Plant TM, Hess DL, et al: On the sites of the negative and positive feedback actions of estradiol in the control of gonadotropin secretion in the rhesus monkey. *Endocrinology* 102:1008–1014, 1978

53. Knobil E, Plant TM, Belchetz PE, et al: Control of the rhesus monkey menstrual cycle: Permissive role of hypothalmic gonadotropin-releasing hormone. *Science* 207:1371–1373, 1980

54. Wildt L, Marshall G, Knobil E: Experimental induction of puberty in the infantile female rhesus monkey. *Science* 207:1373–1375, 1980

55. Lemarchand-Bereaud T, Marie-Madeline Z, Reymond M, et al: Maturation of the hypothalamo-pituitary-ovarian axis in adolescent girls. *J Clin Endorinol Metab* 54:241–246, 1982

56. Illig R, Torresani T, Bucher H, et al: Transient rise in luteinizing hormone and follicle-stimulating hormone secretion during puberty studies in 113 healthy girls with tall stature. *J Clin Endocrinol Metab* 54:192–195, 1982

CHAPTER 3

Ovarian Dysfunction in the Adolescent

Leonore C. Huppert

1. INTRODUCTION

The causes of ovarian dysfunction in the adolescent are fascinating and varied. They may range from the mild oligomenorrhea frequently encountered during the first 1–2 postmenarchal years to complete absence of pubertal development and menstrual function characteristic of the many forms of gonadal dysgenesis. Ovarian dysfunction in the adolescent encompasses the fairly common disorders leading to secondary amenorrhea as well as conditions, such as gonadal dysgenesis, structural anomalies of the female genital tract, and male pseudohermaphroditism, which lead to primary amenorrhea. A brief review of normal sexual differentiation and pubertal development is helpful before considering the numerous ways in which ovarian function may go awry.

2. NORMAL SEXUAL DEVELOPMENT

According to Jost,[1] sexual differentiation may be divided into three stages. The initial stage is the establishment of genetic sex, which is determined at the moment of fertilization. Further normal sexual development depends upon appropriate gonadal differentiation. The final stage is the phenotypic expression of sexual identity, which is determined by the appropriate secretions of the developing gonad.

Leonore C. Huppert • Department of Obstetrics and Gynecology, Medical College of Pennsylvania, Philadelphia, Pennsylvania 19129.

The histologically identifiable step in sexual development is localization of the future germ cells of the new individual. These may be initially identified in the yolk sac outside the embryo about 4 weeks postfertilization. The 1000–2000 germ cells then migrate into the embryo and populate the primitive gonad. A Y chromosome in the developing embryo's karyotype will induce the different gonad to develop into a testis. The intermediary in this induction is felt to be a cell surface antigen known as the H-Y antigen[2] present on all male cells. The absence of the normal Y chromosome and the H-Y antigen will result in differentiation of the primitive gonad into an ovary.

Once the germ cells have populated the ovary, they begin to divide by the process of mitosis. By the eighth week of intrauterine life, the fetal ovaries contain about 500,000 oogonia. These oogonia continue to divide and also start to undergo meiosis, or reductional division, in order to form haploid oocytes. The process of meiosis arrests *in utero* during prophase of the first meiotic division and will not resume until that particular oocyte is being readied for ovulation. At 5 months of intrauterine life, the fetal ovaries contain approximately 7 million germ cells. By birth, however, only about 1 million germ cells remain, the vast majority having degenerated by the process of atresia.

The final expression of gonadal differentiation is the appropriate development of internal and external genitalia. Internal genitalia develop from separate embryologic primordia in the male and the female.[3] In the male, under appropriate stimulation by testosterone, the Wolffian system differentiates into the epididymis, vas deferens, and seminal vesicle.[4] In the female, the mullerian system develops into the fallopian tubes, uterus, cervix, and upper portion of the vagina. Wolffian duct regression in the female takes place without any apparent hormonal stimulation. Mullerian regression in the male, however, requires the specific action of a locally acting peptide hormone produced by the testis known as mullerian-inhibiting factor (MIF).[5]

In contradistinction to the internal genitalia, the external genitalia in the male and female embryo develop from common anlagen. In the male, under the influence of dihydrotestosterone, the urogenitial sinus develops into the prostate and the prostatic urethra.[3,4] Dihydrotestosterone also stimulates the genital tubercle to develop into the glans penis, the genital folds to form the shaft of the penis, and the genital swellings to develop into the scrotum. In the female, on the other hand, the urogenital sinus develops into the urethra and lower portion of the vagina. The clitoris develops from the genital tubercle, and the labia majora and labia minora arise from the genital swelling and genital folds, respectively. It should be noted that normal sexual differentiation in the male is dependent upon three separate hormones: MIF and testosterone, both of which are produced by the fetal testis, and dihydrotestosterone, which is converted peripherally from testosterone by the action of the enzyme 5-α-reductase. Female sexual differentiation, however, does not require any specific hormone stimulation and will take place in the presence of an ovary or in the absence of any gonad.

In addition to the appropriate secretion of hormones by the fetal testes and the appropriate conversion of testosterone to dihydrotestosterone, the male sex hormones must be able to exert their specific actions at the cellular level. It is known that after testosterone enters the target cell it must first bind to a cellular receptor. The hormone–receptor complex then diffuses into the nucleus of the cell where it interacts with the intranucleolar RNA. Dihydrotestosterone must also complex with a cytosol receptor before it, too, can diffuse into the nucleus of the cell and exert its action.[6] Thus, normal sexual development in the male depends not only on appropriate hormone secretion by the testis but also on appropriate conversion of testosterone to dihydrotestosterone in the target cell and on the ability of both testosterone and dihydrotestosterone to complex with cytosol receptor, which must be present in normal amounts and have normal activity.

While gonadal and morphologic sexual differentiation is taking place, development of the stimulatory capacity of the hypothalamus and pituitary is also occurring. By the end of the first trimester of intrauterine life, follicle-stimulating hormone (FSH) and luteinizing hormone (LH) may be detected within the human fetal pituitary gland.[7] Gonadotropin-releasing hormone (GnRH) may also be detected within the fetal hypothalamus in the first trimester.[8] Fetal serum gonadotropin concentrations gradually increase until they reach their peak about 20 weeks of gestation. The midtrimester gonadotropin surge temporally relates to, and may be responsible for, the peak number of germ cells within the fetal ovary. Following the midtrimester peak, gonadotropin levels decline until birth. The decline in fetal gonadotropin levels prior to birth is felt to be the result of maturation of the hypothalamic–pituitary–ovarian negative feedback system, which now becomes very sensitive to the inhibitory control of sex steroids, predominantly of placental origin. The precipitous drop in circulating neonatal estrogen levels, which occurs as a consequence of birth and separation of the placenta, releases the hypothalamic–pituitary axis from negative feedback inhibition and induces a surge of FSH.[9] By 3 months of age, the gonadotropin levels may reach the adult castrate range. The burst of gonadotropin leads to maturation of some ovarian follicles and the elaboration of estrogen.[10] Finally, by 2–4 years of age, the hypothalamus and pituitary have acquired an exquisite sensitivity to the negative feedback effect of estrogen. Thus, the low level of estrogen being produced by the ovary is sufficient to maintain gonadotropin secretion at a basal level throughout the early childhood years.

Between the ages of 8 and 10 years, gonadotropin levels gradually start to rise. The rise is felt to occur as the result of a decreased sensitivity of the hypothalamus and pituitary to the negative feedback effect of the sex steroids.[11] This results in an increased secretion of GnRH, FSH, and LH. The gonadotropins, in turn, then stimulate further estrogen production by the ovary. Ultimately, the ovarian sex steroids and the hypo-thalamic–pituitary axis achieve a new set point of the negative feedback mechanism. It has been estimated that the adult negative feedback system is $\frac{1}{6}$th–$\frac{1}{15}$th less sensitive than the prepubertal system.[11] It is interesting to note that young girls with gonadal dysgenesis also have a rise in go-nadotropins beginning around age 8,[12] thus indicating that an intact gonad is not essential to this prepubertal gonadotropin rise. A unique alteration in the secretory pattern of gonadotropins is another characteristic of the peripubertal years. Marked enhancement of LH secretion occurs during sleep in the late prepubertal and pubertal years.[13] This pattern also occurs in children with gonadal dysgenesis.[14] One of the last steps in the ma-turation of the hypothalamic–pituitary–ovarian axis is the acquisition of positive feedback control by estrogen on gonadotropin release. This is essential for the normal midcycle surge of LH necessary for ovulation. In pre- and early pubertal girls, not only is the ovary unable to produce sufficient estrogen to trigger the LH surge, but it has also been demon-strated that the pituitary lacks the ability to respond to estrogen in a positive fashion.[15]

The development of the secondary sex characteristics in young women is dependent upon secretion of estrogen by the ovaries and an-drogens produced by both the ovary and adrenals. Estrogens are re-sponsible for breast development, alteration of body contour with in-creased fat deposition, accelerated linear growth, skeletal maturation, enlargement of the uterus and cervix, endometrial proliferation, alteration of the vaginal mucosa with increased glycogen deposition, lowering of vaginal pH, and increased vaginal discharge. Androgens are responsible for the growth of pubic and axillary hair and to a lesser degree for the pubertal growth spurt. The earliest sign that puberty is about to begin is the initiation of breast budding. Three additional stages of breast devel-opment have been described by Marshall and Tanner.[16] Shortly after the initial stage of breast development is noted, the first appearance of pubic hair becomes evident. The stages of pubic hair development have also been characterized and described by Marshall and Tanner.[16]

The growth spurt in adolescence is characterized by a period of rapid growth or peak height velocity (PHV) preceded by a stage of lesser growth activity and followed by a stage of decreased velocity and cessation of growth at the time of epiphyseal fusion. The stage of PHV virtually always precedes menarche,[16] hence there is little additional growth in post-menarchal girls. Of all the parameters of pubertal development, the time of menarche is best correlated with bone age. No relationship has been identified between bone age and breast development, pubic hair growth, or PHV. Bone age, together with height and chronologic age, may be used in the prediction of final adult height. Different height tables are available for this calculation. Radiographs of the wrist are most frequently used in assessing bone age.

A recent study of sexual development in American girls demonstrated that the mean age of menarche in the United States is 12.8 years[17] with a normal range from 9.1 to 17.7 years. In the last 80 years, the average age of menarche in western industrialized countries has decreased approximately $2\frac{1}{2}$ years.[18] The decline in the age of menarche has been ascribed to improved socioeconomic conditions and better health care. Genetic factors play a significant role in determining the age of menarche, and there is good correlation between age of menarche of mother and daughter and between sisters.[19]

The menstrual cycles occurring during the first 1–2 years following menarche are usually anovulatory. This has been documented by studying basal body temperature recordings and plasma progesterone levels.[20] It has been estimated that 55–90% of cycles in the first 2 years after menarche are anovulatory and that by 5 years after menarche 20% of cycles still remain anovulatory.[21] This most likely is the consequence of late maturation of the positive feedback effect of estrogen upon the hypothalamus and pituitary.

3. CLASSIFICATION OF MENSTRUAL DYSFUNCTION

Ovarian dysfunction in the adolescent covers a wide spectrum of disorders. These range from mild oligomenorrhea to complete absence of menstruation. The oligomenorrhea exhibited by the adolescent may be a manifestation of an underlying hormonal disorder or, rather, may simply be the result of an immature hypothalamic–pituitary–ovarian axis. Absent menstruation or amenorrhea may be further classified as primary or secondary amenorrhea. Secondary amenorrhea imples that the young girl has previously experienced at least one spontaneous menstrual flow and has subsequently developed amenorrhea. The disorders which can cause secondary amenorrhea in the adolescent are the same as those which are responsible for secondary amenorrhea in the older reproductive-aged woman. The adolescent, in addition, may manifest primary amenorrhea, or failure to initiate menstruation. Disorders leading to primary amenorrhea include all those responsible for secondary amenorrhea and, in addition, include a whole host of developmental and anatomic anomalies, which are uniquely responsible for primary amenorrhea.

Menstrual dysfunction may be the result of failure of any one of the components of the reproductive system.[22] It may be the result of defect within the ovary preventing it from functioning properly. It may also be due to abnormal stimulation of the ovary either by the hypothalamus or pituitary. Finally, the hormonal interrelationships of the reproductive cycle may be perfectly in order, yet there may be an abnormality of the target organ or the outflow tract preventing the normal expression of the cyclic hormonal variation. It is useful to consider the various causes of menstrual dysfunction according to these three separate categories.

3.1. Uterine and Vaginal Anomalies

At the time of the initial pelvic examination of the adolescent with absent menstruation, several rare, but striking, anomalies of the vagina and lower reproductive tract may be recognized. A fairly straightforward diagnosis is that of the imperforate hymen. Typically, the patient is an appropriately developed young woman who has complained of cyclic pain for several months duration. On examination there is a blue bulge at the vaginal introitus as a result of accumulation of menstrual blood within the vagina. Occasionally, the uterus may enlarge as a result of the formation of a hematometra. This condition is easily correctable by a simple surgical procedure.

Of more significance, is absence, atresia, or hypoplasia of the vagina. Embryologically, the upper portion of the vagina forms from the mullerian ducts, while the lower portion of the vagina develops from the urogenital sinus. Failure of development of either the mullerian system or the urogenital sinus or improper fusion of the two will lead to a vaginal anomaly. Congenital absence or atresia of the vagina may also be associated with abnormal uterine development.[23,24] Obviously, the question of whether or not a uterus is present is of vital importance to the patient's childbearing potential. Physical exam and laparoscopy are both useful in this regard. If a uterus is present, the possibility exists that an artificial vagina can be created restoring the potential for fertility. Patients presenting with vaginal anomalies often require counseling to be able to deal psychologically with their problem.[25]

A blind-ending vaginal pouch together with hypoplasia or absence of the tubes and ovaries have been described as the Mayer–Rokitansky–Kuster–Hauser syndrome. This syndrome is due to anomalous mullerian development and has been estimated to occur once in every 4000–5000 female births.[26] When all patients with primary amenorrhea are considered, mullerian agenesis ranks second only to gonadal dysgenesis as a cause of primary amenorrhea.[27,28] Mullerian agenesis is seen significantly more frequently than testicular feminization, which is the third most common diagnosis in primary amenorrhea.[27,28] Since the ovaries are not a part of the mullerian system, patients with the Rokitansky syndrome have normal secondary sexual development and normal ovarian function. Because of the common embryologic origins of the urinary and mullerian systems, there is a high incidence of urologic anomalies in patients with mullerian agenesis.[28] An intravenous pyelogram must, therefore, be an integral part of the evaluation of these patients. Spinal anomalies have also been described in these patients.[29] The karyotype of these patients is always 46XX. Patients with mullerian agenesis must be distinguished from those with testicular feminization, who also present with a blind-ending vaginal pouch and absent uterus. Patients with testicular feminization have a 46XY karyotype and will be discussed below.

Treatment for patients with the Rokitansky syndrome focuses on creating a functional vagina. Fertility is rarely an issue except in those few cases where a functioning uterus is present. Vaginal enlargement can be achieved by means of the Frank dilator technique.[30] Wabrech and associates[31] have found that in 90% of cases this nonoperative technique is satisfactory. The McIndoe procedure is reserved for those patients who do not respond to the dilator technique or in those patients where the potential for fertility exists.[32]

Uterine causes of secondary amenorrhea are quite rare in the adolescent. Basically, uterine causes of secondary amenorrhea are due to a functional failure of the endometrium. The most common underlying cause of endometrial failure is Asherman syndrome. This condition of intrauterine synechiae is found in patients whose basal layer of endometrium was damaged at the time of a pregnancy-related curettage.[33] It may be seen following a pregnancy termination by suction curettage or when curettage is necessary to control postpartum hemorrhage or to evacuate a mole. As pregnancy and pregnancy termination becomes more common in the adolescent, a rise in the frequency of this condition among adolescents may be expected to occur. Asherman syndrome is often suspected after taking a careful history and is further suggested by demonstrating hormonal incompetence of the endometrium. Failure of the patient to withdraw from progesterone alone or from sequential estrogen and progesterone makes the diagnosis of Asherman syndrome a likely possibility. The diagnosis can be confirmed by hysterosalpingography or hysteroscopy. Treatment involves lysis of the adhesions either by dilatation and curettage (D and C) or with the aid of the hysteroscope. Following lysis of the adhesions, any remaining endometrium is stimulated by large doses of conjugated estrogens and progesterone. A small pediatric catheter is inserted in the uterine cavity at the time of lysis of adhesions to prevent the raw side walls of the uterine cavity from readhering. The pediatric foley is left in place for 7 days. In addition to hormonal stimulation, the patient is also treated with a broad-spectrum antibiotic, such as ampicillin, to prevent infection while the foreign body is in place.

Infection is another very rare cause of endometrial destruction. Tuberculosis and schistosomiasis have both been described as causes of endometrial incompetence. The likelihood of finding these diseases of the endometrium in the United States and, moreover, in adolescents is exceedingly unlikely.

3.2. Disorders of the Ovary

Adolescent ovarian failure is principally the result of gonadal dysgenesis and male pseudohermaphroditism. The hallmarks of gonadal dysgenesis are amenorrhea, streak gonads, sexual infantilism, and an abnormal sex chromosome complement. Certain types of gonadal dysgenesis also have associated somatic abnormalities.

The most common form of gonadal dysgenesis is Turner syndrome. Patients with this diagnosis bear a 45XO karyotype. In addition to sexual immaturity and short stature, Turner also described cubitus valgus, webbing of the neck and widely spaced nipples giving the chest a shield-like appearance.[34] Other distinctive features include micrognathia, eipcanthal folds, low-set ears, and ptosis. Coarctation of the aorta has been described in 10–20% of cases. Sixty percent of patients have associated renal abnormalities. A lag in the growth rate of girls with Turner syndrome is often apparent shortly after birth. It is very definitely obvious by the time the patient is ready to enter puberty. It is of interest that even though these children are definitely shorter than would be predicted by parental height, there does appear to be a distinct correlation with parental height suggesting that other genetic factors are still in play.[35]

The typical appearance of these patients together with absent secondary sexual development is highly suggestive of the diagnosis of Turner syndrome. After confirming a hypoestrogenic state by a lack of withdrawal bleeding following a progesterone challenge, determination of gonadotrophin levels establishes gonadal failure. The diagnosis is then confirmed by obtaining a karyotype.

Turner syndrome is the most common cause of primary amenorrhea in young girls.[36] It has been estimated to occur with a frequency of 1 in every 2235 female births.[36] The 45XO karyotype is noted even more frequently in aborted fetuses and may account for up to 20% of spontaneous abortions. It has been estimated that over 99% of conceptuses with a 45XO karyotype will abort. While infertility is the general rule in patients with 45XO Turner syndrome, ten cases of pregnancy in women with this karyotype have been described in the world literature.[37]

There is another group of patients who also present with sexual infantilism, streak gonads, and varying expressions of the typical Turner syndrome phenotype. These are patients with the chromatin-positive variants of gonadal dysgenesis and represent about one-fifth of all patients with gonadal dysgenesis. These patients usually exhibit a mosaic karyotype with 45XO, 46XX being the most common karyotype. These patients are not invariably short and are less likely to exhibit the various associated stigmata of Turner syndrome. Of even more interest is the fact that these patients may menstruate and even bear children.[38] Knowing what we do about the origin of the streak gonad in Turner syndrome, it is possible to understand why the mosaic Turner patient may ovulate and menstruate. In a study of aborted 45XO fetuses ranging in age from 5 weeks to 4 months, Singh and Carr[39] demonstrated that in the early weeks of gestation, primordial germ cells migrate normally into the embryo and populate the gonadal ridge. After the third month of gestation, however, there appear to be an increase in connective tissue and an impaired formation of primordial follicles in 45XO fetuses compared to 46XX fetuses. This suggests that two active X chromosomes are necessary for the normal development of oogonia. It also suggests that in the absence of the second

X chromosome, there is an accelerated rate of atresia of germ cells. Thus, patients with 45XO, 46XX mosaicism, half of whose cells contain a normal sex chromosome complement, can be expected to maintain a certain number of oocytes within their gonads. Yet these patients also bear the 45XO cell line, accounting for an accelerated rate of germ cell atresia. If the gonad of these mosaic patients becomes depleted of germ cells prior to the onset of puberty, these patients will present as primary amenorrhea. If, however, the gonad does not become depleted until the patient is in her thirties, the patient will then present as a case of premature menopause.

Other chromatin-positive variants of gonadal dysgenesis include those patients with structural abnormalities of the second X chromosome. These include patients with deletions or absence of the short or long arms as well as patients with a ring-shaped second X chromosome. It is interesting to note that there is an increased incidence of Hashimoto thyroiditis and diabetes mellitus in patients with structural abnormalities of the X chromosome.[40] These patients also have fewer of the somatic stigmata of Turner syndrome than patients with 45XO karyotype. Studies of patients with varying deletions of the X chromosome suggest that both arms of the X chromosome are necessary for normal gonadogenesis, while the short arm of the X chromosome appears to bear the genetic material that is necessary for normal stature and to prevent the expression of the somatic anomalies associated with Turner syndrome.[41]

Other patients who present with sexual infantilism and gonadal failure have a Y chromosome present in their karyotype. The most common karyotype in this type of disorder is 45XO, 46XY. Differentiation of the gonads may range from bilateral streaks to bilateral dysgenetic testis.[38] The development of the internal and external genitalia correlates with the extent of testicular differentiation. Thus, the phenotype may be completely female, ambiguous, or male. Those patients who have a streak gonad on one side and a testis on the other side are classified as having mixed-gonadal dysgenesis. The importance of obtaining a karyotype in all patients with gonadal dysgenesis must be stressed since it is crucial to establish whether or not a Y chromosome is present in the karyotype. The finding of a Y chromosome in the karyotype of any phenotypic female puts that patient at high risk of developing a malignancy in the gonad. In one series of 42 patients with mixed-gonadal dysgenesis, 9 were found to have tumors at the time of laparotomy.[42]

Much less common variants of gonadal dysgenesis are those patients with structural abnormalities of the Y chromosome. These abnormalities are quite rare and are associated with a variable phenotype depending upon the degree of testicular differentiation. The extent to which these patients manifest the various somatic abnormalities of Turner syndrome also is quite variable and does appear to depend upon which portion of the Y chromosome is deleted.

Not all patients with gonadal dysgenesis have an abnormal karyotype.

A number of patients have been described in the literature who have either a 46XX or a 46YY karyotype associated with bilateral streak gonads, female phenotype, and sexual infantilism and who are of normal stature and who lack the somatic features of Turner syndrome. This disorder has been characterized as pure gonadal dysgenesis.[43] A familial mode of inheritance has been described in a number of cases.[44] As in patients with mixed-gonadal dysgenesis, there is a high risk of malignancy in the streak gonad, usually a gonadoblastoma or dysgerminoma, in those patients bearing a Y chromosome in their karyotype. Thus, prophylactic gonadectomy is indicated as soon as the diagnosis is made.

A fascinating rare entity has been variously described as agonadism, embryonic testicular aggression, or vanishing testis syndrome.[45] The phenotypic appearance of these individuals varies from a female lacking internal genital structures to a normal male with developed Wolffian structures. Whether the patient has a female or male appearance, the karyotype is always 46XY, and sexual infantilism is invariably present. At laparotomy, no gonadal structure can be found. The presumption in these patients is that a testis was present early in development and due to unknown influences be they teratogenic, infective, or genetic, regression of the gonad occurred. Since the early gonad produced MIF, internal female genitalia are not present in these individuals. The stage in embryonic development at which the testicle stops functioning determines the extent of masculinization of the internal genitalia. Obviously those individuals demonstrating a female phenotype experienced a very early loss of testicular function.

For the sake of completeness, Noonan syndrome should also be mentioned.[46] These patients have the typical external appearance of patients with Turner syndrome. Both males and females may be affected. The karyotype is appropriate for the phenotype. Girls with this disorder have functioning ovaries. Puberty may be delayed in these individuals, but normal sexual development will eventually occur. Occasional familial cases have been reported.

The next large group of patients presenting with primary amenorrhea due to abnormal development of the gonad are those with male pseudohermaphroditism. The gonad in these individuals is not an ovary but a testis. Masculinization does not occur appropriately either because of abnormally low testosterone levels due to defective synthesis or because of end organ insensitivity. The term *male pseudohermaphroditism* refers to the fact that the phenotypic appearance of the individual (female) is at variance with the genotype of the individual (XY). Occasionally some forms of male pseudohermaphrodites do present with genital ambiguity. Since this chapter deals with abnormalities in menstruation in young girls, we will consider only those forms of male pseudohermaphroditism that occur primarily in phenotypic females.

The enzymatic defects that affect testosterone synthesis may be subdivided into two categories: The first includes three defects early in biosynthesis of testosterone, which affect both corticosteroid and sex hormone synthesis. The enzymatic defect most likely to produce a normal-looking female is that of the 17-α-hydroxylase deficiency.[47] This enzyme converts pregnenolone to 17-hydroxypregnenolone and progesterone to 17-hydroxyprogesterone. Inhibition of steroid synthesis at this point would prevent synthesis of androgens, estrogens, and corticosteroids. Predictably, these patients are sexually immature and do not initiate puberty. Males who present with this disorder have female external genitalia and male internal genitalia. They, too, fail to develop at puberty. Male and female patients with this disorder are hypertensive, since, as a result of the enzymatic blockage, precursor steriods are converted into mineralocorticoids. This is an exceedingly rare disorder.

The two other early enzymatic defects, the cholesterol desmolase deficiency and the 3-β-hydroxysteroid dehydrogenase deficiency, have such a high infant mortality secondary to the severe adrenal insufficiency that they are rarely noted in adolescents. A few patients with the 3-β-hydroxysteroid have entered puberty, and virilization was noted to occur at that time.[48]

Two additional enzymatic defects occur late in steroidogenesis and affect only the synthesis of the sex steroids. A deficiency in the 17,20-desmolase enzyme, which converts 17-hydroxypregnenolone to dehydroepiandrosterone and 17-hydroxyprogesterone to androstenedione, thus prevents the synthesis of testosterone precursors and ultimately of testosterone also. It has been reported in a phenotypic female with testes and male internal genitalia. Ambiguous genitalia have also been described in these patients.[49] The final defect in sex steroid synthesis is the 17-β-hydroxysteroid oxidoreductase deficiency.[50] This enzyme catalyzes the conversion of androstenedione to testosterone and estrone to estradiol. External genitalia in these patients may be female or ambiguous. At puberty, there is both virilization and breast development along with failure to menstruate.

The final group of patients who present with amenorrhea as a result of a gonadal abnormality are those with the various syndromes of androgen resistance. These patients are genetically normal 46XY males with perfectly normal testes producing normal levels of male hormone. The problem appears to be in the ability of the androgen to enter the target cell and exert its hormone action. When testosterone enters the target cell it is converted to dihydrotestosterone by the enzyme 5-α-reductase. Both testosterone and dihydrotestosterone then complex with a cytosol receptor. The hormone–receptor complex enters the cell nucleus and exerts its biologic action. Testosterone is responsible for stimulation of the Wolffian ducts to form normal internal male genitalia, while dihydrotestosterone is responsible for virilization of external genitalia *in utero* as well as sexual maturation at puberty. Various defects have been described following the entry of testosterone into the target cell, which prevent it from exerting its appropriate biologic action.

The most common form of androgen resistance syndrome is testicular feminization. It is the third most frequent diagnosis made in young girls presenting with primary amenorrhea. It is also the most common form of male pseudohermaphroditism.[51] These patients have a very typical appearance. Although their presenting complaint may be amenorrhea, they are well feminized with good breast development and normal female body contours. Of significance is the absence of normal axillary and pubic hair. Pelvic examination reveals a blind vaginal pouch. When confronted with a well-feminized patient with a blind vaginal pouch and no palpable internal female genitalia, the two diagnoses which must be immediately considered are those of the Rokitansky syndrome and testicular feminization. A karyotype will clearly distinguish between the two individuals.

The defect in testicular feminization is an abnormal androgen receptor. Defective binding as well as an unstable thermolabile receptor have been described in this disorder.[52,53] The hormone profile in testicular feminization reveals normal to slightly elevated plasma testosterone levels, somewhat elevated levels of LH and increased estrogen production by the testis. This increased level of estradiol is responsible for the feminization of these patients at puberty. Because these patients feminize at puberty, castration, indicated due to the XY karyotype, is usually deferred until after puberty. The inheritance of this disorder is X-linked.[54]

Ten percent of patients with testicular feminization present with the incomplete or partial form.[55] They differ from patients with the complete form of the syndrome in that there is ambiguity of the external genitalia during the prepubertal years and both virilization and feminization at puberty. While all mullerian derivatives are absent, there may be some evidence of Wolffian duct differentiation in patients with the incomplete form of the syndrome. Since pubertal virilization occurs in these patients, removal of the gonads is recommended prior to puberty. The underlying defect in incomplete testicular feminization is felt to be abnormal binding of androgen to receptor.

A defect in the enzyme 5-α-reductase limiting conversion of testosterone to dihydrotestosterone has also been described.[56] Since normal levels of testosterone are present throughout fetal and extrauterine life there is normal development of internal male genitalia. Abnormally low levels of dihydrotestosterone prevent normal virilization of the external genitalia during embryogenesis. It is unclear why these patients virilize at puberty since the development of secondary sex characteristics in males is also dependent on dihydrotestosterone. In contrast to patients with testicular feminization, patients with a deficiency in 5-α-reductase do not develop gynecomastia at puberty. The defect appears to be expressed only in males who are homozygous for this autosomal recessive gene.[53]

3.3. Pituitary Disorders

Diseases of the pituitary that prevent the normal elaboration of gonadotropins ultimately lead to abnormal menstruation. The most common pituitary disorder causing amenorrhea or dysfunctional uterine bleeding is the pituitary adenoma. A large percentage of these adenomas are hormonally active. Excessive amounts of growth hormone and adrenocorticotrophic hormone produced by pituitary tumors cause acromegaly and Cushing's disease, respectively. Approximately 70% of pituitary tumors, however, produce prolactin and elevations in this hormone are most frequently associated with menstrual abnormalities. Other space occupying lesions such as gummas or granulomas, may also be responsible for abnormal gonadotropin secretion.

Until a sensitive radioimmunoassay for prolactin was developed, many patients who presented with amenorrhea were never properly diagnosed. It is now known that approximately 20% of patients who present with amenorrhea will have associated elevated prolactin levels.[57] While the finding of galactorrhea increases the likelihood that hyperprolactinemia is present, the absence of galactorrhea certainly does not rule out the possibility of elevated prolactin levels. Asymptomatic adenomas of the pituitary gland are quite frequent as demonstrated by various autopsy series. One autopsy series in 1936 found unsuspected pituitary adenomas present in 22.5% of 1000 cases examined.[58]

Pituitary adenomas are classified as micro- or macroadenomas. Microadenomas are 10 mm or less in diameter. If a pituitary adenoma can be detected on a routine X ray of the sella turcica then a macroadenoma is present. These tumors are felt to be extremely slow growing and may be present for many years before their effects become clinically apparent. Typically, prolactin-secreting adenomas are found in the lateral aspects of the gland. Unilateral blistering and erosion of the sella turcica account for the typical radiologic appearance of a pituitary tumor. A sloping floor is frequently noted on the antero-posterior projection with a double floor evident on the lateral projection. Polytomography of the sella turcica is necessary to diagnose the subtle changes produced by microadenomas. Upward extension of the adenoma is a late and uncommon complication. The new generation of computerized axial tomography (CAT) scanners is useful and has replaced polytonography in many centers in localizing and diagnosing pituitary microadenomas.

Because of the frequency with which pituitary adenomas are diagnosed in patients with amenorrhea, a serum prolactin level should be part of the evaluation of any young woman presenting with markedly irregular or absent menstruation. If the prolactin level is elevated, a CAT scan of the sella turcica should then be obtained. Even if the prolactin level is normal, the possibility of tumor is not completely ruled out. A routine radiographic examination of the skull with coned down views of the sella

turcica can be obtained in these cases to rule out tumors other than pro-
lactinomas. If the radiologic studies are abnormal, the patient should be
referred to an ophthalmologist for visual field evaluation.

The management of patients with pituitary adenomas remains somewhat controversial today since there are no long-term studies comparing
and contrasting the advantages and disadvantages of medical and surgical
management. The following guidelines may be useful, however. If a patient has a large adenoma with or without a disturbance in her visual fields,
surgical removal of the adenoma is recommended. Most pituitary adenomas are amenable to removal by the transsphenoidal approach.[59] In this
procedure the initial incision is made under the lip and extended upward.
The nasal septum is then remove, and entry is made into the sphenoid
sinus. The floor of the sella turcica is then visualized and may be removed
to permit access to the gland. The risks of this procedure are low in terms
of morbidity and mortality. The major postoperative complication is cerebrospinal fluid rhinorrhea, which may affect a very small number of
patients.[60]

It is also accepted that the young patient with a small microadenoma
who is not desirous of childbearing can be treated in a conservative fashion. This patient may be simply observed with periodic reevaluation of
prolactin levels and intermittent radiologic reevaluation of the sella turcica. Frequent polytomography of the sella turcica is contraindicated because of the high dose of radiation received by the optic nerves. Once a
microadenoma is diagnosed, the patient may be followed with routine
skull films. Alternatively, the patient may be treated with bromoergocryptine (Parlodel) a dopamine agonist, highly effective in lowering prolactin levels even in the presence of a tumor. Evidence is now accumu-
lating that treatment with Parlodel may successfully reduce the size of
pituitary tumors as well as return prolactin levels to the normal range.[61]
The patient must be cautioned that there is an 80% likelihood that she
will resume menstruation following treatment with Parlodel and that ovulation will occur. Barrier contraceptive methods are, therefore, appropriate in these patients. Since it is known that estrogen can stimulate the
growth of pituitary adenomas, the oral contraceptive is contraindicated
in patients with this diagnosis. The dilemma in managing patients with
microadenomas of the pituitary arises in those patients who wish to conceive and, therefore, will be encountered less frequently in an adolescent
population. The problem is whether to treat these patients medically or
surgically. The concern is that under the influence of the high estrogen
level during pregnancy, the tumor will enlarge and put the patient at risk
for sudden visual field loss or pituitary apoplexy. If the surgical approach
is chosen, the patient will be without risk to her pituitary during pregnancy, but there is always the concern that if the tumor is large enough,
damage to normal pituitary tissue will occur resulting in panhypopituitarism. On the other hand, medical treatment with Parlodel is quite successful in inducing ovulation and achieving pregnancy. Strict adherence

to the Food and Drug Administration recommedations for use of this drug, however, prohibits using bromoergocryptine for induction of ovulation in patients desirous of pregnancy who have documented pituitary tumors. Many, however, use this drug with great success and have found that the incidence of tumor-related complication during the pregnancy is much lower than expected.[62]

If tumor can be excluded as a present cause of hyperprolactinemia, other diagnoses must be considered. Hyperprolactinemia has been noted in association with primary hypothyroidism. Thyrotrophin-releasing hormone (TRH) is known to stimulate the release of both thyroid-stimulating hormone (TSH) and prolactin. The low levels of thyroid hormone in primary hypothyroidism stimulate the hypothalamus to release TRH resulting in elevation of both TSH and prolactin levels. A simple TSH determination is sufficient to screen for primary hypothyroidism. Enlargement of the sella turcica can occur in women with primary hypothyroidism. The extent of sellar enlargement correlates well with the degree of TSH elevation.[63] Appropriate treatment of these patients with thyroid replacement therapy not only causes the endocrine profile to return to normal but also results in return of the sella turcica to normal dimensions.

Elevated prolactin levels have also been associated with oral contraceptive use and should be suspected in cases of post-pill amenorrhea. Since a number of series of adenomas show a high percentage of antecedent pill usage, some authors have suggested that the estrogen in the oral contraceptive may be responsible for rapid growth of occult microadenomas.[64] An association is by no means certain, however, for many series show no evidence of tumor in patients with post-pill amenorrhea.[65]

A careful history of drug use should be made in any patient with hyperprolactinemia, since a number of drugs are known to elevate prolactin levels. These drugs are those principally interfering with a dopamine action and metabolism. They may be categorized as dopamine receptor blockers (phenothiazines), catechol-depleting agents (reserpine) and α-methyldopa (Aldomet), and those drugs which interfere with the reuptake of dopamine (imipramine, amphetamines).

The empty sella syndrome is a condition that must be distinguished from pituitary adenoma in those patients with a radiologically abnormal sella turcica. This condition, described originally in 1951, is characterized by a congenital defect in the formation of the diaphragma sellae, which permits the subarachnoid fluid to flow into the pituitary fossa. The pituitary gland becomes flattened around the rim of sella turcica. The sella turcica becomes symmetrically enlarged from the pressure of the cerebrospinal fluid. This condition is rare in adolescence, since it is most frequently diagnosed in the middle-aged obese woman.[66] Many patients with this disorder may be hormonally normal although hypogonadotropism has been described. The syndrome of galactorrhea–amenorrhea has also been described in patients with an empty sella.[67]

For the sake of thoroughness Sheehan syndrome should be mentioned. This is the syndrome of postpartum pituitary necrosis resulting from obstetric hemorrhage and shock. Destruction of the gland may be extensive and panhypopituitarism may occur. The onset of the syndrome from the time of original insult is quite variable and may be many years in duration. Since this is a disease that occurs during and after the reproductive years, it will rarely be in the differential diagnosis of the adolescent girl with menstrual dysfunction.

3.4. Hypothalamic Disorders

Hypothalamic dysfunction is considered to be the most common cause of secondary amenorrhea in adolescent and reproductive-aged women. Often it is a diagnosis of exclusion when no other obvious endocrinopathy can be demonstrated. Hypothalamic dysfunction may occur as a result of compression from an extrinsic lesion, a diagnosis particularly pertinent in the young adolescent. In addition, hypothalamic amenorrhea may be the result of a loss of the normal integrative function of the hypothalamic–pituitary–ovarian axis.

The most common extrinsic lesion accounting for hypothalamic dysfunction is the craniopharyngioma. This diagnosis must be considered in any adolescent presenting with evidence of pituitary dysfunction since craniopharyngiomas typically present during the second decade of life.[68] These tumors are felt to arise from remnants of Rathke's pouch and are usually present along the anterior surface of the pituitary stalk. The tumors may be cystic, solid, or a combination of both. Headache, growth retardation, and delayed sexual maturation are common presenting complaints. Approximately 70% of patients with craniopharyngiomas have visual impairment ranging from total blindness to optic atrophy with bitemperal hemianopsia being the most common defect.[69] Endocrinologically, growth hormone and the gonadotropins are the most frequently affected hormones.[70] X rays reveal enlargment of the pituitary fossa and suprasellar calcification. Surgical decompression is often necessary. Because of the location of the tumor, complete resection is often impossible and surgery must be followed by radiotherapy. Other hypothalamic neoplasms which may result in abnormal function include the germinoma, also known as ectopic pinealoma, and the very malignant endodermal-sinus tumor. Gliomas of the optic nerve may also lead to compression of the hypothalamic region.

Infective and infiltrative diseases may also lead to hypothalamic dysfunction. Hand–Schüller–Christian (histiocytosis X) may result in infiltration of the hypothalamic region with eosinophilic granulomas. This disorder may lead to delayed puberty secondary to hypogonadatropic hypogonadism, growth retardation, and diabetes insipidus in children.[71] Sarcoidosis may also cause hypothalamic destruction. Since growth retardation is often the presenting complaint in patients with these disorders, the pediatrician rather than the gynecologist is often consulted first.

A quite rare and fascinating syndrome of hypogonadotropic hypogonadism associated with anosmia has been described. When originally described by Kallman, the disorder was felt to be a hereditary one limited to males. The anosmia is the result of congenital agenesis of the olfactory aparatus. More recent studies have demonstrated that this condition may also exist in women.[72] These patients present with primary amenorrhea and minimal to moderate secondary sexual development. LH and FSH levels may be quite low but in most cases will respond to GnRH.[73] Careful testing of the patient's impaired olfaction is essential to making the diagnosis. Hormone replacement therapy is appropriate in these patients until fertility is desired. Ovulation induction can then be accomplished with human menopausal gonadotrophins, since clomiphene citrate is ineffective in the absence of adequate gonadotrophin reserves.

By far the largest number of adolescent patients presenting with irregular menses or secondary amenorrhea are considered to have a functional form of hypothalamic anovulation. In the very young patient, this may well represent immaturity of the hypothalamic–pituitary–ovarian axis as was discussed in the introductory section of this chapter. In addition, stress and extremes of weight may also be related to anovulation. Frisch and co-workers[74,75] have set forth the interesting concept of a critical weight necessary for the onset of normal menstrual function and also for the restoration of menstrual cyclicity in cases of secondary amenorrhea due to weight loss. Lachelin and Yen[77] studied 11 patients with psychogenic amenorrhea, 3 of whom were adolescents. They established that the pituitary trophic hormones (LH, FSH, TSH, prolactin, and growth hormone) as well as the ovarian hormones (estrogens and androgens) were within the normal range for the early follicular portion of the menstrual cycle. The basic levels of the system appear to be intact in this disorder with the primary defect being that of loss of cyclic function. A very large number of these patients resumed spontaneous menstruation on their own.

Perhaps at the extreme of psychogenic amenorrhea is the amenorrhea associated with anorexia nervosa. A review of the endocrine abnormalities in this disorder indicate that it is truly a separate disease, however, and does not merely represent one end of the continuum of hypothalamic amenorrhea. These patients have a typical psychological profile which includes a hyperactive, aggressive, overachieving personality, loss of perception of their own bodily image, and self-starvation to the point of emaciation, illness and death.[77] Circulating gonadotropin levels are low with LH suppression greater than that of FSH.[78] Various studies of the gonadotropin response to GnRH in this condition have indicated a delayed or blunted response.[78,79] Even more fascinating is the finding that the gonadotropin pattern reverts to the peripubertal pattern of elevated LH levels during sleep.[80] The abnormal responses to GnRH disappear as weight is gradually regained.[81] These alterations in GnRH responsiveness and in gonadotropin secretory patterns are not seen in other patients with hypothalamic amenorrhea. Treatment of this disorder is directed to the underlying psychiatric problems. Resumption of menstruation will occur when recovery is underway and weight is gradually regained.

Post-pill amenorrhea has also been ascribed to hypothalamic dysfunction. Whether post-pill amenorrhea is a discreet entity is open to debate. Many gynecologists feel that a large number of patients experiencing post-pill amenorrhea were those who experienced menstrual irregularity prior to taking the oral contraceptive.[82] The incidence of post-pill amenorrhea appears to be approximately equal to the incidence of spontaneous amenorrhea of greater than 6 months duration in the general population.[83] Before assuming the amenorrhea is due to the ingestion of the oral contraceptive, a thorough evaluation for other etiologies, including pituitary adenoma, should be carried out.

Somewhat defying compartmentalization is the type of chronic anovulation typified by the polycystic ovary syndrome. The clinical presentation of patients with this syndrome may be quite varied.[84] In addition to amenorrhea or dysfunctional uterine bleeding, hirsutism and obesity are frequently observed. It should be noted, however, that neither hirsutism nor obesity are essential for making the diagnosis of the polycystic ovary syndrome. The basal hormone profile and hormone dynamics in this disorder distinguish it from chronic hypothalamic anovulation. LH levels are tonically elevated in patients with the polycystic ovary syndrome. FSH levels, on the other hand, are in the low to low normal range. The elevated LH levels are felt to be the result of an exaggerated pulsatile response to GnRH.[85] It is known that elevated estrogen levels can increase the sensitivity of the LH response to GnRH, and this mechanism has been implicated in these patients who have abundant circulating estrogen. The predominant estrogen in these patients is estrone rather than estradiol and is formed in the peripheral tissues principally from the conversion of androstenedione.

The polycystic ovary syndrome is also characterized by the inappropriate secretion of ovarian hormones. Both testosterone and its precursors including dehydroepiandrosterone, dehydroepiandrosterone sulfate, and androstenedione are elevated. In some patients the testosterone levels are noted to be in the normal range.[86] Free testosterone levels, however, are elevated, and the daily production rate of testosterone is increased. Normal testosterone levels are maintained by relatively low levels of sex hormone-binding globulin and an increased metabolic clearance for testosterone. Ovarian androgens are produced by the theca interna under the stimulation of LH. The androgen precursors then diffuse into the avascular granulosa where they are converted into estrogen by the action of an FSH-dependent aromatizing enzyme.[87] Patients with polycystic ovaries are noted to be deficient in their ability to aromatize androgen precursors, which may account for the elevated androgen levels and the depressed estradiol values in these patients.[88]

Histologically, the polycystic ovary is characterized by numerous small subcapsular follicular cysts. The capsule of the ovary is smooth and thickened as in any ovary that does not ovulate. The appearance of these ovaries is felt to be the consequence rather than the cause of the anovulation. Because the gross appearance of the polycystic ovary yields no clues to its diagnosis, the laparoscope has no role in the diagnosis of this syndrome.

The follicles in the polycystic ovary represent various early stages in the maturation process. This failure of normal follicular maturation may be the result of the low normal FSH levels present in this condition or the failure of these ovaries to produce adequate amounts of estrogen which serves to augment the action of FSH on the maturing follicle by increasing FSH receptors.[89] Follicular maturation may also be inhibited by the elevated levels of ovarian androgens, which are felt to block various stages in follicular growth causing follicular atresia.[90] In summary, Yen[91] has proposed that the polycystic ovary syndrome develops as a result of chronic inappropriate estrogen feedback upon the hypothalamic–pituitary–ovarian axis. Excess ovarian androgen production results in the extraglandular conversion of these androgens to estrogen. The elevated estrone levels, in turn, increase pituitary sensitivity to GnRH resulting in an exagerated pulsatile release of LH. The elevated circulating levels of LH then further stimulate the ovarian thecal cells to produce even more ovarian androgen. The decreased FSH levels seen in this syndrome together with the increased ovarian androgen lead to an inhibition of follicular maturation and decreased ovarian estrogen production. The system, therefore, operates in an acyclic steady-state condition. The trigger phenomenom that initiates this condition is not yet known.

Treatment of this disorder depends on the desire of the patient. If the patient is a young adolescent concerned about her cosmetic appearance, an attempt should be made to decrease the level of androgen in hopes that her hirsutism will improve. This can best be accomplished by means of the combined oral contraceptive.[92] Since the low-dose oral contraceptive is effective in suppressing the activity of the hypothalamic–pituitary system, there is no reason why it cannot be used in these patients. The efficacy of this form of therapy can be evaluated by the lack of further progression of the hirsutism and by demonstrating suppression of the abnormal androgen values into the normal range. If the patient does not exhibit hirsutism but is amenorrheic, treatment should still be instituted since these patients are at risk for endometrial hyperplasia. Provera®, 10 mg, given the first to the fifth day of each or every other month will induce a withdrawal flow and protect the endometrium from hyperplasia. Dysfunctional uterine bleeding will also be prevented. In the patient desirous of pregnancy, clomiphene citrate is the drug of choice. Because these patients are well estrogenized, they are ideal candidates for clomiphene treatment and respond with a high rate of ovulation.[85] The gratifying clinical response to clomiphene in patients with the polycystic ovary syndrome makes wedge resection of the ovaries a rarely indicated procedure. Because of the risk of development of peritubular adhesions as a result of the surgery, it should be reserved for those patients who are clomiphene failures.

4. DIAGNOSIS

Obviously, an accurate diagnosis of ovarian dysfunction in the adolescent rests upon a detailed history and a careful physical examination. The history should include information on significant medical illnesses, a drug history, and a chronological summary of the patient's growth and development of secondary sex characteristics. The patient should also be questioned about her current weight and significant past fluctuations in weight. If the patient has menstruated, an attempt should be made to determine whether the bleeding episodes were ovulatory or anovulatory. Pertinent questions should try to establish the regularity of bleeding, the length of menstrual flow, the heaviness of menstrual flow, and the presence or absence of symptoms associated with ovulation, such as dysmenorrhea, breast tenderness, backache, and premenstrual mood changes. The menstrual history of the mother and sisters may also be helpful. Physical examination should not be limited to a pelvic examination but should include a general evaluation. The degree of sexual maturity including breast development and amount of axillary and pubic hair should be noted. If gonadal dysgenesis is suspected, a search should be made for the typical somatic stigmata. Examination of the external genitalia will reveal whether there is any evidence of ambiguity or virilization. Any deviation from the normal female hair pattern should be noted. An imperforate hymen or blind vaginal pouch will be easily detected. If the internal and external genitalia appear to be anatomically normal, an impression should be formed concerning the extent to which the genitalia are being stimulated by estrogen. A vaginal smear from the midportion of the lateral vaginal wall can be evaluated for hormone effect. A small hypoplastic uterus and cervix are frequently noted in hypoestrogenic young girls. Good estrogen levels manifest themselves by the appearance of copious mucus in the cervical os. If a sample of this mucus is allowed to air-dry on a glass slide and is then examined under a microscope a typical fern pattern can be recognized in well-estrogenized patients.

The initial phase of the diagnostic evaluation is also directed at determining the extent to which the patient is estrogenized. This can easily be determined by administering a progesterone challenge in the form of 10 mg Provera, given daily for 5 days. If sufficient estrogen to cause endometrial proliferation is present, the progesterone will induce a withdrawal flow. Obviously, pregnancy should be ruled out before the Provera is administered. A withdrawal flow following progesterone administration not only gives proof of endogenous estrogen production but also indicates that a responsive endometrium with a normal outflow tract is present. Endogenous estrogen also indicates that the ovary, pituitary, and hypothalamus are functioning to a certain extent. If withdrawal bleeding fails to occur following a progesterone challenge, a distinction must be made between a hypoestrogenic state and an unresponsive endometrium. Since an unresponsive endometrium is very unusual in the adolescent girl, the more likely diagnosis is that of estrogen deficiency. This can be documented by demonstrating that the patient will bleed in response to an estrogen and progesterone challenge. If estrogen levels are deficient, the problem can rest at either the ovarian level or in improper stimulation of the ovary by the hypothalamus or pituitary. Ovarian failure can easily be separated from hypothalamic and pituitary dysfunction on the basis of gonadotropin levels. Elevated gonadotropin levels in the menopausal range are indicative of ovarian failure. In the adolescent presenting with primary amenorrhea and ovarian failure, a karyotype is then in order. A buccal smear is not sufficient to determine whether or not a Y chromosome is present. As previously mentioned, those patients with a Y chromosome in their karyotypes require removal of the gonad because of its malignant potential.

All patients who present with amenorrhea should have their prolactin levels determined. If the prolactin levels are elevated, a TSH level should also be obtained to rule out primary hypothyroidism. Elevated prolactin levels require evaluation of the sella turcica by means of a CAT scan. Visual field determinations are ordered if the CAT scan is abnormal. If the prolactin level is normal, a routine skull X ray with coned down views of the sella turcica is helpful to rule out any abnormality of the sella turcica other than that associated with hyperprolactinemia. Additional hormone tests are obtained as indicated. Signs and symptoms of hyper- or hypothyroidism should prompt an evaluation of thyroid function. Hirsutism can be evaluated by means of a serum testosterone and a 24-hour urine collection for 17-ketosteroid determination. Dehydroepiandrosterone sulfate may be used as a screening test in place of the urinary 17-ketosteroid determination.[93] If the testosterone or the 17-ketosteroid value is elevated, further investigation of the origin of these excess androgens can be performed according to the schema of Speroff et al.[21]

5. THERAPY

Clearly, the treatment of young girls with ovulatory dysfunction varies according to the diagnosis. Individual treatment for specific disorders have been discussed earlier in the chapter. This section will deal only with the treatment of abnormal bleeding or amenorrhea in the absence of demonstrable pathology.

The anovulatory cycles of young girls in the first several years following the menarche are often characterized by episodes of irregular and frequently profuse or prolonged bleeding. This is the result of breakthrough bleeding from an estrogen-primed endometrium. When endometrium is exposed solely to estrogen, as is the case in anovulatory states, the tissue often becomes overstimulated and outgrows its hormonal support. At that point, irregular shedding occurs. Treatment of dysfunctional bleeding is directed at providing adequate hormonal support so that healing will take place and bleeding will cease. This may best be accomplished with the combined oral contraceptive given in high doses. The key to this regimen is that both estrogen and progesterone are being supplied, which will stimulate growth of a structurally stable endometrial lining. Once pregnancy has been satisfactorily rule out, the patient should be given a 21-day pack of 35- or 50-μg estrogen pills and instructed to take one pill four times a day until they are all gone. Bleeding should stop within 1–2 days. Occasionally, the patient may complain of nausea as a result of the high estrogen dose. The patient should also be advised that upon discontinuing the pills she will experience another episode of vaginal bleeding, but this one will be limited in duration and flow.

Occasionally, if the irregular bleeding has been very light, a course of 10 mg, Provera, daily for 5 days will be sufficient to stop the bleeding and induce a withdrawal flow. Progesterone works by converting the excessively stimulated estrogen-primed endometrium into structurally stable secretory endometrium, which will shed thoroughly on discontinuing the medication. If the patient's bleeding has been heavy, however, there will be insufficient estrogen-primed endometrium remaining upon which the progesterone can exert its action, and the progesterone will be ineffective. In these cases, it is far wiser to begin treatment immediately with a combined regimen.

Rarely, bleeding is heavy enough to require hospitalization. In these cases an attempt should be made to stop the bleeding with 25 mg Premarin® intravenously every 4 hr until the bleeding starts to slacken. At that point the patient may be started on the high-dose oral contraceptive regimen. It should be noted that D and C is never a first line of therapy for a patient with dysfunctional uterine bleeding. The disorder has an underlying hormonal defect and responds well to hormonal therapy. D and C is reserved for those cases failing to respond to hormonal management suggesting other pathological diagnosis.

Once the initial bleeding episode is under control, further bleeding should be controlled on a regular every-other-month basis with 10 mg Provera given the first to the fifth day of the month. With this treatment plan, the patient is being supplied with the hormone she is lacking, and additional episodes of irregular bleeding will be prevented. In addition, the endometrium will be protected from hyperplastic changes. Progesterone alone is preferable to the combined oral contraceptive unless contraception is desired, since progesterone will not suppress the hypothalamic–pituitary axis and will permit the patient to resume spontaneous menses when her system has matured. Progesterone should also be used on a periodic basis to induce withdrawal bleeding in those amenorrheic patients who respond.

The hypoestrogenic patient including those with gonadal dysgenesis and hypothalamic amenorrhea should be treated with monthly estrogen and progesterone therapy. Estrogen stimulation is important for development of the secondary sex characteristics and for the long-term prevention of osteoporosis. Premarin, 0.625mg, should be administered on the 1st to the 25th day of each month together with 10 mg Provera on Days 16–25. A withdrawal flow will then occur at the end of each month. Patients with gonadal dysgenesis should, of course, be treated continuously. Patients with hypothalamic amenorrhea need to be reevaluated on a periodic basis as well as given the opportunity to allow their own reproductive systems to mature, so they should be treated on an intermittent basis.

REFERENCES

1. Jost A: A new look at the mechanisms controlling sex differentiation in mammals. *Johns Hopkins Med J* 130:38–53, 1972
2. Ohno S: The role of H-Y antigen in primary sex determination. *J Am Med Assoc* 239:217–220, 1978
3. Wilson JD: Sexual differentiation. *Annu Rev Physiol* 40:279–306, 1978
4. Siiteri PK, Wilson JD: Testosterone formation and metabolism during male sexual differentiation in the human embryo. *J Clin Endocrinol Metab* 38:113–125, 1974
5. Picard J-Y, Tran D, Josso N: Biosynthesis of anti-mullerian hormone by fetal testes: Evidence for the glycoprotein nature of the hormone and for its disulfide-bonded structure. *Mol Cell Endocrinol* 12:17–30, 1978
6. Higgins SJ, Gehring U: Molecular mechanisms of steroid hormone action. *Adv Cancer Res* 28:313–397, 1978
7. Siler-Khodr TM, Morgenstern LL, Greenwood FC: Hormone synthesis and release from human fetal adenohyophysis in vitro. *J Clin Endocrinol Metab* 39:891–905, 1974
8. Kaplan SL, Grumbach MM, Aubert ML: The antogenesis of pituitary hormones and hypothalamic factors in the human fetus: Maturation of central nervous system regulation of anterior pituitary function. *Recent Prog Horm Res* 32:161–243, 1976
9. Winter JSD, Faiman C, Hobson WC, et al: Pituitary gonadal relations in infancy. I. Patterns of serum gonadotropin concentrations from birth to four years of age in man and chimpanzee. *J Clin Endocrinol Metab* 40:545–551, 1975
10. Winter JSD, Hughes IA, Reyes FI, et al: Pituitary gonadal relations in infancy. II. Patterns of serum gonadal steroid concentrations in man from birth to two years of age. *J Clin Endocrinol Metab* 42:679–686, 1976
11. Grumbach MM, Roth JC, Kaplan SL, et al: Hypothalamic-pituitary regulation of puberty

in man: Evidence and concepts derived from clinical research, in Grumbach, MM, Grave, GD, Mayer, FE (eds): *Control of the Onset of Puberty*. New York, Wiley, 1974, pp 115–166

12. Conte FA, Grumbach MM, Kaplan SL: A diphasic pattern of gonadotropin secretion in patients with the syndrome of gonadal dysgenesis. *J Clin Endocrinol Metab* 40:670–674 1975

13. Boyar R, Finkelstein J, Roffwarg H, et al: Synchronization of augmented luteinizing hormone secretion with sleep during puberty. *N Engl J Med* 287:582–586, 1972

14. Boyar RM, Finkelstein JW, Roffwarg H, et al: Twenty-four hour luteinizing hormone and follicle stimulating hormone secretory patterns in gonadal dysgenesis. *J Clin Endocrinol Metab* 37:521–525, 1973

15. Reiter EO, Kuhn HE, Hamwood SM: The absence of positive feedback between estrogen and luteinizing hormone in sexually immature girls. *Pediatr Res* 8:740–745, 1974

16. Marshall WA, Tanner JM: Variations in the pattern of pubertal changes in girls. *Arch Dis Child* 44:291–303, 1969

17. Zacharias L, Rand WM, Wurtman RJ: A prospective study of sexual development and growth in American girls: The statistics of menarche. *Obstet Gynecol Survey* 31:325–337, 1976

18. Tanner JM: Trend towards earlier menarche in London, Oslo, Cophenhagen, the Netherlands and Hungary. *Nature (London)* 243:95–96, 1973

19. Zacharias L, Wurtman RJ: Ageat menarche. *N Engl J Med* 280:868–875, 1969

20. Winter JSD, Faiman C: Pituitary-gonadal relations in female children and adolescents. *Pediatr Res* 7:948–953, 1973

21. Apter D, Vihko R: Serum pregnenolone, progesterone, 17-hydroxyprogesterone, testosterone, and 5-α-dihydrotestosterone during female puberty. *J Clin Endocrinol Metab* 45:1039–1048, 1977

22. Speroff L, Glass RH, Kase NG: *Clinical Gynecologic Endocrinology and Infertility*, ed 2. Baltimore, Williams & Wilkins, 1978

23. Akinkugbe A: Vaginal atresia and crytomenorrhea. *Obstet Gynecol* 46:317–319, 1975

24. Griffin JE, Edwards C, Madden JD, et al: Congenital absence of the vagina, the Mayer-Rokitansky-Kucter-Hauser syndrome. *Ann Int Med* 85:224–236, 1976

25. David A, Carmil D, Bar-David E, et al: Congenital absence of the vagina, Clinical and psychologic aspects. *Obstet Gynecol* 46:407–409, 1925

26. Bryan AL, Nigro JA, Counseller VS: One hundred cases of congenital absence of the vagina. *Surg Gynecol Obstet* 88:79–86, 1946

27. Ross GT, VandeWiele RL: The ovaries, in Williams, RH (ed): *Textbook of Endocrinology*, ed 5. Philadelphia, Saunders, 1974, p 368

28. Fore SR, Hammond CB, Parker RT, et al: Urologic and genital anomalies in patients with congenital absence of the vagina. *Obstet Gynecol* 46:410–416, 1975

29. Turunen A, Unnérus C-E: Spinal changes in patients with congenital aplasia of the vagina. *Acta Obstet Gynecol Scand* 46:99–106, 1967

30. Frank RT: The formation of an artifical vagina without operation. *Am J Obstet Gynecol* 35:1053–1055, 1938

31. Wabrek AJ, Millard PR, Wilson WB, Jr, et al: Creation of a neovagina by the Frank nonoperative method. *Obstet Gynecol* 37:408–413, 1971

32. McIndoe AH, Banister JB: An operation for the cure of congenital absence of the vagina. *J Obstet Gynaecol Br Commonw* 45:490–494, 1938

33. Klein SM, Garcia CR: Asherman's syndrome: A critique and current review. *Fertil Steril* 24:722–735, 1973

34. Turner, HH: A syndrome of infantilism, congenital webbed neck and cubitus valgus. *Endocrinology* 23:566–574, 1938

35. Sarto GE: Cytogenetics of fifty patients with primary amenorrhea. *Am J Obstet Gynecol* 119:14–23, 1974

36. Jacobs PA: The incidence and etiology of sex chromosone abnormalities in man. *Birth Defects* 15:3–14, 1979

37. Wray HL, Freeman MVR, Ming PL: Pregnancy in the Turner's syndrome with only 46X chromosomal constitution. *Fertil Steril* 35:509–514, 1981
38. Morishima A, Grumbach MM: The interrelationship of sex chromosome constitution and phenotype in the syndrome of gonadal dysgenesis and its variants. *Ann NY Academy Sci* 155:695–715, 1968
39. Singh RP, Carr DH: The anatomy and histology of XO human embryos and fetuses. *Anat Rec* 155:369–383, 1967
40. Williams ED, Engel E: Thyroiditis and gonadal dysgenesis. *N Engl J Med* 270:805–810, 1964
41. Grumbach MM, VanWyk JJ: Disorders of sexual differentiation, in Williams, RH (ed): *Textbook of Endocrinology*, ed 5. Philadelphia, Saunders, 1974, p 462
42. Federman DD: *Abnormal Sexual Differentiation*. Philadelphia, Saunders, 1967, p 68
43. Harnden DG, Stewart JSS: The chromosomes in a case of pure gonadal dysgenesis. *Br Med J* 2:1285–1287, 1959
44. Simpson JL: Genetic aspects of gynecologic disorders occurring in 46, XX individuals. *Clin Obstet Gynecol* 15:157–182, 1972
45. Edmon ED, Winters AJ, Porter JC, et al: Embryonic testicular regression, a clinical spectrum of XY agonodal individuals. *Obstet Gynecol* 42:208–214, 1976
46. Noonan JA: Hypertelorism with Turner Phenotype. *Am J Dis Child* 116:373–380, 1968
47. Biglieri EG, Herron MA, Brust N: 17-Hydroxylation deficiency in man. *J Clin Invest* 45:1946–1954, 1966
48. Opitz JM, Simpson JL: Pseudovaginal perineoscrotal hypospadias. *Clin Genet* 3:1–26, 1972
49. Zachman M, Völlmin JA, Hamilton W, et al: Steroid 17,20-desmolase deficiency: A new cause of male pseudohermaphroditism. *Clin Endocrinol* 1:369–385, 1962
50. Saez JR, Morena AM, DePeretti E, et al: Further *in vivo* studies in male pseudohermaphroditism with gynecomastia due to a testicular 17-ketosteroid reductase defect (Compared to a case of testicular feminization). *J Clin Endocrinol Metab* 34:598–600, 1972
51. Jaqiello G, Atwell JD: Prevelence of testicular feminization. *Lancet* 1:329, 1962
52. Keenan BS, Meyer WJ III, Hadjian AJ, et al: Syndrome of androgen insensitivity in man: Absence of 5-α-dihydrotestosterone binding protein in skin fibroblasts. *J Clin Endocrinol Metab* 38:1143–1146, 1974
53. Griffen JE: Testicular feminization associated with a thermolabile androgen receptor in cultured human fibroblasts. *J Clin Invest* 64:1624–1631, 1979
54. Meyer WJ III, Migeon BR, Migeon CJ: Locus on human X chromosome for dihydrotestosterone receptor and androgen insensitivity. *Proc Nat Acad Sci USA* 72:1469–1472, 1975
55. Griffen JE, Wilson JD: The syndrome of androgen resistance. *N Engl J Med* 302:198–209, 1980
56. Imperato-McGinley J, Guerrero L, Gautier T, et al: Steroid 5-α-reductase deficiency in man: An inherited form of male pseudohermaphroditism. *Science* 186:1213–1215, 1974
57. Franks S, Murray MAF, Jequier AM, et al: Incidence and significance of hyperprolactinemia in women with amenorrhea. *Clin Endocrinol* 4:597–607, 1975
58. Costello RT: Subclinical adenoma of the pituitary gland. *Am J Pathol* 12:205–216, 1936
59. Hardy J: Transphenoidal hypophysectomy. *J Neurosurg* 34:581–594, 1971
60. Post KD, Biller BJ, Adelman LS: Selective transphenoidal adenectomy in women with galactorreha amenorrhea. *JAMA* 242:158–162, 1979
61. McGregor AM, Scanlon MF, Hall K: Reduction in size of a pituitary tumor by bromocrepine therapy. *N Engl J Med* 300:291–293, 1979
62. Gemzell C, Wang CF: Outcome of Pregnancy in women with pituitary adenoma. *Fertil Steril* 31:363–372, 1979
63. Yamada T, Tsukuit, Ikejiri K, et al: Volume of sella turcia in normal subjects and in patients with primary hypothyroidism and hyperthyroidism. *J Clin Endocrinol Metab* 42:817–822, 1976

64. Sherman BM, Harris CE, Schlechte J, et al: Pathogenesis of prolactin secreting adenomas. *Lancet* 2:1019–1021, 1978

65. Kleinberg DK, Noel GL, Frantz AG: Galactorrhea: A study of 235 cases including 48 with pituitary tumors. *N Engl J Med* 296:589–600, 1977

66. Hodgeson SE, Randall RV, Holman CB, et al: Empty sella syndrome. *Med Clin N Am* 56:897–907, 1972

67. Check JH, Rakoff AE, Goldfarb A, et al: Amenorrhea-galactorrhea associated with the empty sella syndrome. *Am J Obstet Gynecol* 128:688–689, 1977

68. Banna M: Craniopharyngioma: Based on 160 cases. *Br J Radiol* 49:206–233, 1976

69. Kennedy B, Smith RJS: Eye sings in craniopharyngioma. *Br J Ophalmol* 59:689–695, 1975

70. Jenkins JS, Gilbert CJ, Ang V: Hypothalamic-pituitary function in patients with craniopharyngiomas. *J Clin Endocrinol Metab* 43:394–399, 1976

71. Strauss JH, Yen SSC, Benirschke JH, et al: Hypothalamic hypopituitarism in an adolescent girl: Assessment by a direct functional test of the adenohypophysis. *J Clin Endocrinol Metab* 39:639–644, 1974

72. Tagatz G, Fialkow PJ, Smith D, et al: Hypogonadotropic hypogonadism associated with anosmia in the female. *N Engl J Med* 283:1326–1329, 1970

73. Oettinger M, Bruneteau DW, Psaroudakis A: FSH and LH response to LHRF in Kallman's syndrome. *Obstet Gynecol* 47:233–236, 1976

74. Frisch RE: Weight at menarche: similarity for well-nourished and undernourished girls at differing ages and evidence for historial constancy. *Pediatrics* 50:445–450, 1972

75. Frisch RE, McArthur JW: Menstrual cycles: Fatness as a determinant of minimum weight for height necessary for their maintenance or onset. *Science* 185:949–951, 1974

76. Lachelin GCL, Yen SSC: Hypothalamic chronic anovulation. *Am J Obstet Gynecol* 130:825–831, 1978

77. Bruch H: Perceptual and conceptual disturbances of anorexia nervosa. *Psychoso Med* 24:187–194, 1962

78. Vigersky RA, Loriaux DL, Anderson AE, et al: Delayed pituitary hormone response to LRF and TRF in patients with anorexia nervosa and with secondary amenorrhea associated with simple weight loss. *J Clin Endocrinol Metab* 43:893–900, 1976

79. Beumont PJV, George GCW, Pimstone BL, et al: Body weight and the pituitary response to hypothalamic releasing hormones in patients with anorexia nervosa. *J Clin Endocrinol Metab* 43:487–496, 1976

80. Boyar RM, Katz J, Finkelstein JW, et al: Anorexia nervosa: Immaturity of the 24-hour luteinizing hormone secretory pattern. *N Engl J Med* 291:861–865, 1974

81. Warren MP, Jewelewicz R, Durenforth R, et al: The significance of weight loss in the evaluation of pituitary response of LH-RH in women with secondary amenorrhea. *J Clin Endocrinol Metab* 40:601–611, 1975

82. Higgins GR: Contraceptive use and subsequent fertility. *Fertil Steril* 28:603–612, 1977

83. Pettersson F, Fries H, Nillius JS: Epidemiology of secondary amenorrhea. *Am J Obstet Gynecol* 117:80–86, 1973

84. Goldzieher JW, Green JA: The polycystic ovary. I. Clinical and histology features. *J Clin Endocrinol Metab* 22:325–331, 1962

85. Rebar R, Judd HL, Yen SSC, et al: Characterization of the inappropriate gonadotropin secretion in polycystic ovary syndrome. *J Clin Invest* 57:1320–1329, 1976

86. Kirschner MA, Zucker IR, Jespersen D: Idiopathic hirsutism: An ovarian abnormality. *N Engl J Med* 294:637–640, 1976

87. Armstrong DT, Papkoff H: Stimulation of aromatization of exogenous and endogenous androgens in ovaries of hypophysectomized rats *in vivo* by follicle-stimulating hormone. *Endocrinology* 99:1144–1151, 1976

88. Spaeth DG, Osawa Y: Estrogen biosynthesis. III. Stereospecificity of aromatization by normal and diseased human ovaries. *J Clin Endocrinol Metab* 38:783–786, 1974

89. Richards JS, Midgley AR, Jr: Protein hormone action: A key to understanding ovarian follicular and luteal cell development. *Biol Reprod* 14:82–94, 1976

90. Louvet J-P, Harman SM, Schreiber JR, et al: Evidence for a role of androgen in follicular maturation. *Endocrinology* 97:366–372, 1975
91. Yen SSC: Chronic anovulation due to inappropriate feedback system, in Yen SSC, Jaffe RB (eds): *Reproductive Endocrinology*, ed 2, Philadelphia, Saunders, 1978, pp 297–323
92. Givens JR, Anderson RN, Wiser WL, et al: Dynamics of suppression and recovery of plasma FSH, LH, androstenedione and testosterone in polycystic ovarian disease using an oral contraceptive. *J Clin Endocrinol Metab* 38:727–735, 1974
93. Korth-Schutz S, Levine LS, Neu MI: Dehydroepi-androsterone sulfate (DS) levels, a rapid test for abnormal adrenal androgen secretion *J Clin Endocrinol Metab* 42:1005–1013, 1976

CHAPTER 4

Sexual Behavior in Adolescence

Harold I. Lief

1. INTRODUCTION

If the shortest and post parsimonious definition of mental health is the capacity to love and to work, adolescence is the period of life during which those capacities are being forged. If young adulthood provides the opportunity for the fine tuning of these capacities and skills, the gross tuning of these capacities takes place during adolescence. The ability to love includes being able to function fully as a sexual being. The issues surrounding this developmental aspect of personality are addressed in this chapter, as are the special problems in medical care created by these issues.

Sexual pressures, internal and external, create a good deal of the discomfort for the adolescent. Likewise, adolescent sexual problems are apt to cause feelings of discomfort in the physician, at least until he or she has acquired sufficient experience. It is, however, a time of opportunity for both adolescent and physician; the adolescent has the opportunity to forge his or her own personality in a healthy direction, and the physician has the chance to intervene effectively and to practice preventive medicine as well as provide therapy.

Definition of Terms

The title of the chapter indicates an emphasis on behavior, yet it is impossible to discuss behavior without taking into account attitudes and

Harold I. Lief • Department of Psychiatry, Pennsylvania Hospital, Philadelphia, Pennsylvania 19106; and School of Medicine, University of Pennsylvania, Philadelphia, Pennsylvania 19104.

feelings. Attitudes and feelings are not only precursers of behavior but are consequences of it as well. In discussing feelings, it is important to differentiate between erotic sensations and emotions, for example, anxiety over dating or guilt over masturbation. For that reason, I will avoid the term *feelings* and talk about either bodily sensations or emotions.

Attitudes are a complex of beliefs, values (preferences), and emotions. For example, many religions inculcate negative attitudes toward premarital sex. This would include the belief that premarital sex is sinful, the feelings of guilt if one engages in it, or even if one is tempted to engage in it, and the value or preference for chastity. In discussing adolescent sexuality, one must also take into account the presence of conflict. The adolescent may have an internal conflict between erotic desires, pushing him or her toward sexual experimentation, and the belief that such experimentation is sinful. There may also be an interpersonal conflict between his or her desire for sexual abstinence and peer pressure to engage in sex or a conflict between the adolescent's wish to have certain sexual experiences that violate the wishes of parents.

Other terms need defining, for example, puberty. Puberty is the biological surge of maturation that creates the capacity to reproduce as well as the change in primary and secondary sexual characteristics that result in an adult appearance. Somewhat arbitrarily, its midpoint is defined as menarche in girls and seminal emmission in boys. Adolescence is the period of psychologic and social response to puberty; there are wide variations cross-culturally and among individuals.

Adolescence is also arbitrarily separated into early, middle, and late adolescence, usually categorized as early adolescence from 13 to 15, middle adolescence from 15 to 17, and late adolescence from 17 to 19. Clearly, bodily changes, emotions, attitudes, and behavior are strikingly different in early adolescence and late adolescence.

2. THE SEXUAL SYSTEM

Physical sex is only one part of the totality of one's sexual being, better referred to as sexuality. One's sense of being male and masculine, female and feminine, and the various roles these self-perceptions engender or influence, are important ingredients of sexuality. Cognitive, emotional, and physical sex all contribute to the totality. In this sense, our sexuality is what we *are* rather than what we *do*. Sexuality may be described in terms of a system, somewhat analogous to the circulatory or respiratory system. The components of the sexual system are set forth in Table I.[1]

TABLE I. The Sexual System

Biological sex—chromosomes, hormones, primary and secondary characteristics
Core gender identity (sometimes called *sexual identity*—sense of maleness and femaleness)
Gender identity—sense of masculinity and femininity
Gender role behavior, including behavior motivated by a desire for sexual pleasure, ultimately orgasm (physical sex), as well as gender behavior with masculine and feminine connotations

---

2.1. Biological Sex

Ordinarily, the chromosomal pattern, either XX or XY, determines the phenotype. There is usually synchrony between the chromosomal pattern and the hormonal programming of the brain, notably the hypothalamus, during fetal life, depending upon the presence or absence of androgen. In the male, fetal testicular androgens are secreted between the 6th and 12th week of fetal life leading to the normal development pattern found in the vast majority of male infants. Similarly, the relative absence of androgens in the XX fetus will lead to the development of female characteristics. Abnormal chromosomal patterns, such as those that occur in Klinefelter or Turner syndrome, or androgen insensitivity due to some abnormality of the receptor sites in the male, create problems of intersexuality to be discussed later in this chapter.

2.2. Core Gender Identity

Some of the intersex patients, and some without evident biologic defects, have grave difficulty in developing a core gender identity consistent with their biologic sex. These are transsexuals. If they are biologic males, they think of themselves as a female trapped in a male body. Transsexual biologic females consider themselves to be a male trapped in a female body. Patients with these and other sexual problems involving core gender identity frequently are labeled as cases of gender dysphoria.

2.3. Gender Identity

With these relatively rare exceptions, development of sexuality leads to a secure sense of maleness or femaleness, which is generally complete by the age of 3. However, in our culture, doubts about masculinity and femininity are ubiquitous, and these often are particularly troublesome during adolescence. This topic will be discussed under Section 8.

2.4. Gender Role Behavior

Gender role behavior refers to those behaviors that encompass a whole spectrum of roles for a particular sex, male or female, in a particular cultural system. The culture has orientations as to, for example, whether that sex should place primary emphasis on an occupational career or on a family or should develop attributes of aggressiveness or of nurturance. The culture defines in sometimes clear and sometimes not so clear terms the expectations of society for males and females and the expectations of the males and females in each institution in society. There is a reciprocal relationship between gender identity and gender role behavior. How one feels about oneself as either masculine or feminine affects one's behavior in institutional settings; conversely, one's behavior affects the way one feels about one's sex.

Schwartz et al.[2] divide sexual role behavior into *proceptive*, *acceptive*, and *conceptive* phases. The proceptive is the phase of initial erotic behaviors involving attraction, solicitation, and courtship. In the absence of a partner, it may be initiated in imagery and fantasy. The subjective experience includes anticipatory sexual drive, interest, or desire. The proceptive phase may be viewed in its longitudinal aspects as part of psychosexual development. It may be seen as part of a single sexual encounter or of an ongoing relationship.

Proception is followed by acception, the phase of bodily interaction and potential genital union. This phase subdivides into stages of excitement, plateau, orgasm, and resolution. Excitement involves a subjective sense of sexual pleasure accompanying physiologic changes, primarily those of vasocongestion. Plateau consists of sustained sexual excitement and progressive physiologic changes. Orgasm consists of a subjective sense of ecstatic peaking of sexual pleasure with which sexual tension is resolved. In both sexes, orgasm involves spasmodic contractions, in the male the prostate, seminal vesicles, and urethra, and in the female the outer third of the vaginal vault. Usually orgasm is also accompanied by general muscular contractions, including involuntary pelvic thrusting. Resolution consists of general muscular relaxation and a sense of well-being and of contentment.

Sexual inhibition may occur at the proceptive stage leading to inhibition of dating and courtship or anxiety-ridden or neurotic forms of pair bonding. Inhibition in the acceptive phase leads to the specific dysfunctions that will be described in Section 8.

The conceptive phase is the phase of conception, pregnancy, childbirth, and eventually parenthood. Pregnancy and parenthood create special problems in adolescence. Problems involving the conceptive or reproductive phase, including contraception, pregnancy counseling, and abortion, are discussed in Chapters 5 and 6.

3. ASPECTS OF PHYSICIAN–PATIENT INTERACTION

3.1. Physician Comfort

The adolescent presents many problems in patient care, not the least of which is the sometimes awkward and uncomfortable interaction between physician and teenage patient. The patient is often frightened, perhaps angry, and often rather uncommunicative. The physician is sometimes at a loss about how to decrease these negative emotions and defenses to develop rapport sufficient for adequate communication. If the physician is relatively inexperienced in dealing with the emotional aspects of adolescence, the contagiousness of the patient's discomfort is augmented by the physician's anxiety about his or her competence.

Discomfort may be enhanced if there are ethical conflicts between physician and adolescent patient. If the physician's own values are threatened by the adolescent's sexual behavior, behavior which the physician may regard as immoral, unaesthetic, or undisciplined, the physician may respond with negative or angry reactions. If the physician has not learned, either in medical school or in his graduate training, to be comfortable with a wide variety of sexual behaviors and to realize the great range of normalcy in sexual matters, he or she may be markedly uncomfortable. Some physicians, for example, cannot recommend contraception to sexually active adolescent females, even for those who have already had an unplanned pregnancy. Even more physicians are unable to become involved in problem-pregnancy counseling, if abortion is one of the options. Physician comfort and a nonjudgmental attitude toward a wide range of behaviors that the physician may not countenance for himself or for his own children, is a requirement of effective physician–adolescent patient interaction when sexual concerns are being discussed. The physician who cannot fill this role appropriately should face the issue squarely by indicating his discomfort, or lack of experience, and refer the patient to a colleague. In making the referral, the physician should avoid any implications of rejection or humiliation. Changing social and sexual mores and their implications for patient management will be discussed in the next section.

Some physicians are uncomfortable in initiating discussions of sexual concerns. Since the adolescent is often reluctant to discuss sexual matters openly, it is up to the physician to initiate such discussion. We will return to this issue in Section 6.

3.2. Roles of Physician

Not infrequently there is a conflict between two physician roles, between that of listener–inquirer, eliciting pertinent information, and the authoritarian pseudoparental role often played by physicians, either deliberately or stylistically. If the physician comes across as a sort of parent or, at any rate, takes an authoritarian stance, as a consequence of anticipated disapproval, the adolescent may become noncommunicative.

3.3. Confidentiality

Of great concern is the issue of confidentiality. While the assurance of confidentiality is absolutely necessary in order to elicit the often grudging trust of the adolescent, there are some booby traps of which the physician must be aware. If the adolescent, for example, has a venereal disease, in most instances it is necessary for the physician to report this to an appropriate public health agency. It is possible, especially in early or midadolescence, that some sexual behavior may have serious legal or ethical consequences, and the parents may have to be informed. It is almost always a matter of clinical judgment, but where confidentiality is threatened by these or similar situations, the physician's task is to be very open and to let the adolescent know the dilemma the physician faces and ask for cooperation in revealing the relevant information in an appropriate manner. Ethical issues occasionally occur, which are indeed perplexing. For example, suppose a family physician or a pediatrician had under his or her care a young woman and a young man engaged to be married and knew that one of them had an untreated venereal disease or that one of them was a preferential homosexual and that this information had not been shared with the partner. While these instances are rather dramatic, enough cases of this sort occur so that whereas, in the vast majority of instances, confidentiality must be protected at all costs, exceptions are possible.

4. CHANGING SOCIAL AND SEXUAL MORES

Premarital Coitus

The most striking change in sexual behavior in the United States is the increase in premarital coitus.[3-5] However, because of the incorrect stereotype that premarital intercourse was uncommon until recent decades, the change in premarital coital frequencies may be exaggerated. The records, for example, of one Massachusetts church 200 years ago show that one-third of all women there had confessed fornication to the minister (without the confessions, their babies would not have been baptized). Since there were other "virgins" who did not become pregnant and therefore did not have to confess, the number undoubtedly was even greater. This misconception is one of degree, for there was a rise in the incidence of premarital coitus during the 1920s, and another great shift toward earlier coital experience during the late 1960s and early 1970s. Changes in both eras have been more evident among female than among males.

In 1976, slightly more than half of all white women were nonvirgins by age 19. Blacks had a higher incidence of premarital coitus than whites, began coitus earlier, and hence were far more vulnerable to pregnancy. Men tended, as in the past, to start coitus earlier than women and to have more frequent and varied experiences. The majority of men are nonvirgins by their late teens.

The increase in nonmarital coitus during adolescence has created a shift in attitudes. Recent estimates show that the trend toward greater acceptance of premarital coitus, noted by Kinsey in the 1940s, has accelerated since then. It is not now considered as socially deviant as it was a decade or so ago. By no means does this indicate that adolescents are invariably psychologically or socially prepared for coitus, as the higher rates of teenage pregnancy and sexually transmitted disease (STD) prove. The shift in sexual behavior during adolescence cannot be attributed to the pill. It is now more than 20 years since the pill became available, and sexually active teenagers have never used oral contraceptives regularly, many of them not at all. The availability of contraceptives does not alter the incidence of premarital coitus, although it does seem to aid in containing the risk of pregnancy.

Social class differences do not affect the incidence of coitus as much as they do the use of contraception and abortion. Lower class women are more likely to bear children out of wedlock, probably because contraception and abortion are less accessible to them, and perhaps also less acceptable. This is the major difference between college-bound youth and those that do not attend college.[6] The acceptance of premarital intercourse has affected all strata of our society. Moreover, the double standard for the sexes is disappearing, at least among our youth. The change in attitudes about premarital intercourse has brought about a shift in the decision-making process. Many more young women are less concerned about whether to participate in coitus than was the case a generation ago. They are more concerned about the nature of their relationships. Decisions often involve not whether the young woman should have sex, but with whom. Of course, the shift is relative since a great many young women are still faced with the decision about participation. To a lesser extent, young men also have to go through sometimes painful decisions about "how far to go." The decision to participate may be influenced by increasing peer pressure as well as by the media portrayal of what appears to be freely available sex. Powerful influences, albeit indirect, are brought to bear on sexual decision making.

The increase in premarital coitus might lead the physician to label such behavior as "promiscuous." In the vast majority of instances, this labeling would be incorrect and unfortunate. Studies demonstrate that one-third of men and one-half of women have had no more than one or two partners premaritally. In studies that use the past 6 months as a retrospective time frame, 80% of sexually active young women and 60–70% of sexually active young men have had intercourse with only one partner. It is safe then to say that, if promiscuity is defined as casual and multiple sexual partnerships, most young people are not promiscuous.

The phrase "permissiveness with affection"[7] still remains the norm for young people. Affection, if not love, is an important ingredient for most young women, and for many young men as well. If we label sexual relations as *procreational* (motivated primarily for reproduction), *relational* (the emphasis is on the partnership), and *recreational* (the emphasis is on fun and pleasure), it is true that more young people than ever before are engaged in recreational sex. Nonetheless, the norm is still relational sex.

Young people have far more permissive attitudes toward premarital sex than their elders. Young people seem to be more responsive to the changing social standards, whereas, because of the fear of their teenage children becoming pregnant or contracting a STD, parents of teenagers have become more conservative. An old quip, namely that a conservative is a liberal with a teenage daughter, contains a great deal of truth.

5. PSYCHOSEXUAL DEVELOPMENT IN ADOLESCENCE

Psychosexual behavior is influenced by biologic, psychologic, and social variables. One cannot thoroughly understand any aspect of human behavior without reference to these three dimensions. This is particularly true in adolescence during which rapid changes in hormones and bodily characteristics find their expression in the well-known vacillation in adolescent behavior.

5.1. Biologic Development in the Adolescent Female

Puberty occurs about 2 years earlier in girls than in boys, and the bodily changes characteristic of this period may take from $1\frac{1}{2}$–4 years. There is an increase in gonadotrophins at the time of puberty, and the ovaries produce estrogen in increasing amounts. Estrogen accounts for the development of breasts and uterus as well as the fat distribution that creates the typical female body contours. Androgens, mainly adrenal androstenedione, are also produced in the female and are responsible for the development of pubic and axillary hair and increased growth of the clitoris and labia majora.

Androgen seems to be responsible for sexual desire in both sexes. It certainly seems to be true in older women. The exact relationship between androgen and increase in sexual interest in adolescent girls is unclear; although the percentage of women who experience orgasm increases with age, there does not seem to be any correlation between orgasmic capacity and androgen levels. In women in their twenties, however, androgen levels and sexual arousal, as well as coital frequency, are correlated.[8]

The most important biologic phenomenon in female reproductive physiology is, however, the cyclic regulation of sex hormones under the influence of hypothalamic-releasing hormones. This seems to occur about the middle of puberty somewhere between the ages of 12 and 13 on the average. It is at this time that menarche occurs. There is often a period of 1–2 years of sterility between the onset of menarche and the onset of reproductive capacity. Although this may be nature's way of protecting the female from early pregnancy, obviously there are exceptions. Every teenage pregnancy service in the country reports pregnancies in very young girls, sometimes as early as 9 years of age.

A positive correlation between the early onset of puberty and the frequency of coital orgasm has been reported by Raboch and Bartak.[9] In a study of 1756 patients, they found that the highest frequency occurred in those women in whom menarche took place at age 11 or younger and that the lowest orgasmic frequency correlated with those women in whom menarche appeared at 16 years.

5.2. Biologic Development in the Adolescent Male

Testicles begin to enlarge at about age 12 and produce increasing amounts of testosterone about a year later. This initiates growth of the penis, pubic hair, and prostate, deepens the voice, and causes characteristic male musculature and bone growth. Ejaculatory ability is achieved shortly before age 14 on the average.

Orgasm and ejaculation, while usually integrated in the adult male, are separated in preadolescence or early adolescence as indeed they are also late in life. The major component of orgasm is rhythmic muscular contractions at 0.8-sec intervals. Ejaculation in the adult male ordinarily is superimposed on this basic orgasmic pattern. In adolescent boys, the first ejaculation may occur after a number of orgasmic experiences. The first ejaculation does not occur from nocturnal emission as is commonly thought but much more often occurs through masturbation. Masturbation accounts for two-thirds of first ejaculations, whereas nocturnal emission only accounts for about 12%.

5.3. Psychologic Development in Adolescent Males and Females

Adolescents, both males and females, have the psychologic task of developing a firm gender identity, a sense of masculinity or femininity. Presumably, they have passed through an earlier stage during the first 3 years of life in which core gender identity, or the feelings of maleness or femaleness, have been thoroughly established. They now have the developmental task of developing a sense of security about their gender identity. Anxieties about this are ubiquitous. It is a very rare adolescent indeed who does not feel awkward about his or her initial efforts at exploring relationships with the opposite sex. Fears of embarrassment, of being humiliated or rejected, are universal. To complicate this psychologic task, the upsurge of hormones in early adolescence brings back, at least at unconscious levels, thoughts and feelings (including both erotic sensations and emotions) about the parent of the opposite sex, a resurgence of Oedipal feelings.

The adolescent's gender identity has profound influence on sexual behavior, and sexual behavior reciprocally has a profound influence on gender identity. It is only after gender identity has been fairly well consolidated that true intimacy with others becomes possible. It is through sexual behavior that the young person learns much about interpersonal relations. Complicating gender identity, an essential ingredient in developing the capacity to love, are other psychologic tasks. These are independence from family, especially from parents, occupational goals and identity, or the capacity to work, and the capacity to form friendships, primarily through peer acceptance. It seems as if adolescents, as an essential ingredient in the whole process of separation and individuation, require conflict with the social values of their elders who are promulgating those values; this process seems to take place in every culture.

6. INTERVIEWING AND HISTORY TAKING

Confidentiality, which has been discussed earlier, is not the only issue of significance in the sexual interviewing of adolescents. The attitude of the adolescent has a profound effect on the interview. Frequently the adolescent is suspicious and distrustful of adults, especially those whom he may regard as parental surrogates. He may be ashamed of his sexual feelings or behaviors and fearful of disapproval. He may find it very difficult indeed to even label or identify his feelings. All of these factors affect the style in which the adolescent communicates.[10]

6.1. Communication

As a Group for the Advancement of Psychiatry report[11] states

> It is naive to expect the adolescent patient to be reasonable, open and co-
> operative during the period when he is undergoing rapid and apparently un-
> predictable mental and physical changes. These changes require the adolescent
> to distance himself from the parents and all the other adults he associates with
> them. Moreover, he must justify or rationalize behavior, prompted by his
> search for identity, his fear of his own drive toward independence, his fear of
> sexual impulses toward his parents, and the threat of engulfment he senses in
> being close to a strong adult. A view of the adult (physician) as untrustworthy
> allows him to justify his discomfort that he is being victimized by someone
> hostile and treacherous.

By taking his time, imparting the feeling that he is concerned, and asking questions in such a way that the adolescent feels as if the physician is attempting to understand his feelings and is not being judgmental, the physician begins to establish empathy. Once the adolescent has the feeling that the interviewer is on the same wavelength with him, he will begin to open up and some of the initial distrust and suspicion will decrease. It may take considerable time, however, to establish a relationship which allows for the adolescent to talk freely and comfortably. If the physician feels that the adolescent is being evasive or that the subject is too much to handle, the interview should be kept relatively short and the adolescent given an opportunity to return in a short time to continue the discussion. On the other hand, if the situation demands some prompt action, the physician has to abandon this more leisurely approach and be more con-frontational.

6.2. Other Defenses

The adolescent's anxieties may lead to a show of bravado about sexual matters, perhaps an attempt to impress the interviewer with the patient's sophistication or lack of concern for social decorum. The interviewer must not be fooled by the patient's outward dress and manner, for under the air of bravado and unconcern may lie much anxiety and embarrassment. If there is guilt about sexual impulses or behavior, the teenage patient may blame others or express anger in order to ward off guilty feelings. The physician must try to be patient and relatively imperturbable as he searches for the underlying causes of the adolescent's defensive interview behavior. Since the adolescent may be quite provocative, this is not always easy. The interviewer often finds himself in a delicate situation. In facing the adolescent's suspicion and distrust, he cannot afford to deal with this distrust by indirection or a gradual approach to the issue. He has to face the sexual problem head-on. Any vacillation invites increased hostility and increases the barrier of distrust between the patient and the physician at the very outset.

6.3. Content of Sexual Concerns

The interview will be shaped by the content or substance of the adolescent's sexual concerns. These may be fears of pregnancy, venereal disease, homosexual inclinations, or, perhaps in later adolescence, marriage itself. There may be a fear of closeness, anxieties about sexual performance, guilty fears about sex, and the fear of psychic or physical damage (the latter fear is usually repressed) as a consequence of sexual activity. The adolescent may be engaged in a series of self-destructive sexual acts or relationships. Each of the sexual concerns or behaviors has its own special content, and the interviewer's skill is partially dependent on the extent of his knowledge of the subject under discussion.

6.4. History Taking

The general principles in sexual history taking are listed in Table II.

6.4.1. Comfort

The importance of the physician's comfort has been mentioned earlier. If the physician is ill at ease or anxious, he will communicate these feelings to the patient and his own discomfort will increase proportionately. Conversely, the physician who is comfortable will put his patient at ease and will be able to elicit information far more appropriately and readily.

TABLE II. General Principles in Sexual History Taking

1. Interviewer should be comfortable (feel at ease)
2. Establish empathy
3. Interviewer's values should not impinge
4. Knowledge and skill correlated
5. Precise questioning indicated, with tact
6. Approach emotionally charged areas by degree
7. Proceed from learning, to attitudes, to actual behavior
8. Reassure patient by making statement regarding the common occurrence of a particular behavior

6.4.2. Empathy

The initial goal of the physician should be to establish empathy. Trust and confidence are built up slowly, especially with adolescents, but the beginnings of a therapeutic alliance can occur in the first few minutes of an interview by the therapist. Along with warmth, compassion, and understanding, quickly identifying and labeling the patient's feelings is one of the best ways to start building empathy.

6.4.3. Values

The interviewer must be careful not to allow his own values to needlessly impinge on the interview and history taking. His awareness of his own sexual values should enable him to prevent their unwarranted instrusion into the interview. There are times when the physician may want to share his value position with the adolescent patient, but this decision must be made deliberately, and the words chosen carefully. The last thing the physician wants to do is to create the impression that he is an authoritarian and judgmental parental surrogate.

6.4.4. Knowledge

The greater the therapist's knowledge, the more skillful the history taking, unless discomfort or prejudice interferes. Knowledge expands the repertoire of possible questions, increasing the depth of understanding, and permits the introduction of issues not previously explored. The more the physician knows, the easier it is for him to follow the leads and cues the patient presents and to make new connections for the patient to consider.

6.4.5. Precision

Questions should be as exact and precise as possible within the limits imposed by tact. One should, in general, avoid using street language since this may be seen as phony "togetherness." Street language may be used if the adolescent uses it first, and if he or she does not easily understand technical language.

6.4.6. Bridging

Use bridging questions, going from less highly charged questions to those with more potential impact, e.g., from menarche, to dating, to light petting, to heavy petting, to intercourse; from bodily concerns, to nudity, to masturbation; from wet dreams to masturbation.

6.4.7. Steps

Proceed from inquiring about how the patient learned about certain sexual behaviors, such as homosexuality, to attitudes toward that issue, and then to actual behavior: "At what age did you first learn that some people are attracted physically to members of the same sex?" "What were your attitudes about that?" "What are your current attitudes?" "What kind of homosexual experiences, if any, have you had?"

6.4.8. Consensuality

The patient is often reassured by the physician's statement that a certain kind of sexual behavior is very common. In adolescence, this would include concerns about such behavioral items as masturbation, wet dreams, homosexual impulses or contacts, premarital coitus, or even some aspect of the sexual repertoire, such as oral genital sex. Validating that these behaviors are common prevents the adolescent from thinking that he or she is abnormal, or even weird. The wide range of normal sexual behavior would place most sexual activities in this frame of reference although, to be sure, there are some behaviors, such as paraphiliac behavior, where this type of reassurance would be inappropriate.

7. PHYSICAL EXAMINATION

The sexual connotations of the physical examination tend to be embarrassing for teenagers. Their concerns about their body image, e.g., penis size, breast size and shape, even irrational fears that the physical examination will reveal sexual activity, are conducive to shame and embarrassment. The physician has to emphasize those aspects that are significant in the examination of an adult, namely, treatment of the patient with dignity, insuring the patient's privacy in undressing and dressing, and the supply of gowns and drapes as needed. The physician should be certain to use the adolescent's preferred name instead of an impersonal one, such as "big boy" or "honey child." If the physician is a male, the adolescent's growing adult status should also be demonstrated by the use of a nurse or assistant as a chaperone while examining breasts or doing a pelvic examination. When a female physician is examining the genitalia of a male patient, that potentially embarrassing procedure should be prefaced by an explanation of the purpose of the examination.

The Pelvic Examination

As Litt and Cohen[12] point out, a pelvic examination is indicated if the adolescent is sexually active, if contraceptives might be prescribed, if the patient has any symptom which might be gynecologic in nature, such as discharge, abnormal bleeding, or abdominal pain, when there is a history of diethylstilbestrol exposure during pregnancy, and when the teenager is curious about her body. It is generally recommended that examination of internal genitalia be included in the examination of all females from early childhood as a part of learning about one's body, as well as to facilitate comfort with the pelvic examination in later years. Models of the internal and external anatomy are often very helpful. With adolescent girls, most authorities recommend performing the bimanual examination first and the speculum examination, if indicated, second. For virginal patients, the Pederson or Graves speculum is recommended.

The examination of genitalia in both males and females may give the physician another opportunity to ask about the teenager's sexual concerns. It certainly does provide an opportunity for reassuring the teenage about anatomic normalcy.

8. COMMON SEXUAL PROBLEMS

It is helpful to divide sexual problems into three major categories—*concerns*, *difficulties*, and *disorders*. Concerns deal with a person's anxieties about his or her own sexuality. Difficulties are those problems created by conflict between partners. Disorders include the established syndromes, such as gender dysphorias, paraphilias, and sexual dysfunctions.[13] Using this perspective as a frame of reference, this section of the chapter is devoted to these major topics—concerns of adolescents, concerns of parents of adolescents, sexual dysfunctions, homosexuality, STDs, and sexual aggression, including rape and incest.

8.1. Sexual Concerns of Adolescents

Some of the most common include those relating to body image, nocturnal emissions, masturbation, dating, and decision making.

8.1.1. Body Image Concerns

Concerns by adolescent males about penis size are ubiquitous. Boys measure their masculinity by comparing the size of their phallus with those of their peers in locker rooms and shower rooms. They fail to recognize that the manner in which they perceive their own genitalia, looking straight down at their own phallus, especially when the pubertal fat pad appears in early adolescence, makes it appear as if their own organ is smaller than those they observe by the lateral view of their peers. This illusion may never be corrected for many years, if at all. Second, they fail to recognize that any differences in size during the flaccid state is almost invariably neutralized by the process of erection; that is, the differences are decreased in the erect state. Third, they also fail to realize that the depth of the average vagina is only 4.5 inches deep, and fourth, that the vagina is a potential space enabling it to grasp (in the plateau stage of excitement) or to expand in relation to the size of the object within it. The elimination of these myths and misconceptions will do much to reassure the teenager.

The adolescent male, especially in early adolescence, may easily be concerned over breast enlargement present in at least one-third, and probably in one-half or more, of adolescent males. Reassuring the patient that this is extremely common is necessary and that the gynecomastia will disappear within at most 2 years of its initial appearance. Presumably, the gynecomastia is a consequence of adrenally derived estrogens in puberty.

Perhaps the most frequent of all concerns of the adolescent male is a consequence of acne. The shame over facial appearance may severely inhibit his dating behavior. Treatment of acne may be the most effective "sex therapy" of all!

Adolescent girls are also frequently embarrassed about their breast size or shape. Only on rare occasions is cosmetic breast surgery indicated because of insufficient or excessive size or asymmetry. (Of course, supernumerary nipples should be removed as early as possible). If breast surgery is indicated, it should be delayed until a sufficient stage of maturation is reached; the maturational process may obscure the eventual size of breasts. It is rare that such surgery is recommended until the girl is at least 15 years.

8.1.2. Maturation

Pediatricians report that the most common concern of adolescents is over perceived inadequacy in development of secondary sex characteristics. Litt and Cohen[12] report that "the normal teenagers," attending Stanford University's Youth Clinic, responses to a health concern inventory indicate that 70% harbor such concern." The majority of these are about penis and breast size discussed in Section 8.1.1. Others, however, are concerned about delayed menarche or in the delay of appearance of secondary sex characteristics. It is rather uncommon for delayed sexual maturation in girls to cause serious problems. Boys, however, are disadvantaged by their lack of muscle development and short stature. This creates feelings of inadequacy in athletic prowess as compared to boys of the same age. Boys with delayed maturation may have little facial hair growth and have a high-pitched voice, subjecting them to a lot of teasing by their peers. This usually adversely affects dating behavior. A number of studies have reported long-term disadvantages, such as lower levels of occupational attainment, marriage at later age, fewer children, and less financial success than early-maturing boys.

Delayed puberty in both boys and girls often creates body image problems and a decrease in self-esteem. This, in turn, may lead to difficulties in school performance and may affect other areas of behavior as well.

Delayed puberty is most often constitutional but other causes include endocrine disorders affecting the hypothalamic–pituitary–gonadal axis, hypothyroidism, growth hormone deficiency, chromosomal disorders, and severe chronic illness.

Precocious puberty is fairly rare and more often appears in females. It has been reported that children as young as 2 have started to secrete gonadal hormones, initiating a normal sequence of pubertal changes at a very early age. Girls with this condition may undergo breast development and begin menstruating by the time they are 6–8. While these girls engage in masturbation and have sexual fantasies at an earlier age than normal children of the same age, they do not engage in sexual intercourse earlier than their peers.

In instances of either delayed or precocious maturation, isolation from peers is a problem that has to be managed by appropriate parental and professional support. In developing a treatment plan, the relative absence of peer support must be considered.

8.1.3. Nocturnal Emissions

Until the adolescent boy has been reassured that wet dreams are perfectly normal, he is apt to be anxious, believing them to be evidence of abnormality. Even after he has learned that wet dreams are normal, if he has previously associated them with masturbation, and masturbation with sinfulness, there may be a conditioned guilty reaction to nocturnal emissions. It is not uncommon for the adolescent boy to think that his mother or father will attribute the stained sheets or pajamas to masturbation. It is probably unusual for the adolescent to bring this concern to the physician. He is more apt to discuss it with his peers or, if he is lucky, with an understanding parent or teacher. Nevertheless, the physician should be aware that this may be a source of concern to the early adolescent boy. It is appropriate to ask, "What concerns do you have about wet dreams?" and to make certain that all concerns disappear.

8.1.4. Masturbation

Studies demonstrate that only 3% of males and slightly over 20% of females never masturbate. Its normalcy is not in question; it is a healthy part of sexual development. Despite that, there are still injunctions handed down by society, especially affecting those who have received an orthodox religious upbringing. Even in the absence of explicit proscriptions against masturbation, since early genital caressing and touching is done in secrecy and is often associated with impulses and fantasies that have to be repressed, masturbatory guilt is extremely common. At times, the guilt may be attached to the fantasies accompanying masturbation rather than to the masturbation itself, although this distinction is not commonly made by the adolescent himself. The physician's job is to find out why the adolescent is concerned and to reassure him of its normalcy and the absence of any untoward consequences for his health or development.

During adolescence, girls are probably somewhat more vulnerable to feelings of guilt than are males. Since group discussions are less frequent than among males, their ability to learn to consensually validate the normalcy of their sexual behavior is less. A tactful approach to this problem usually is possible by asking when the adolescent first learned about masturbation, and then asking "How *young* were you when you first started?" The physician should not assume that sex education classes in school will have resolved the problem of masturbatory guilt.

8.1.5. Dating Concerns

Dating usually follows a fairly set cultural pattern: partying with one's own sex, then gathering in groups of heterosexual couples, eventually going out in foursomes, finally with one partner. In ordinary circumstances, this progression relieves enough anxiety and encompasses enough social learning so that most adolescents are fairly comfortable with dating. A large number of them, however, remain anxious about the whole process of searching out and contacting members of the opposite sex. This "sniffing" operation may create intense anxiety. Almost every clinician has come across the adolescent male who is fearful of approaching a desirable female or who gets panicky when he has to call a girl he hopes to date. The fear of rejection, and hence humiliation, is widespread. This is commonly based on poor self-esteem, although it may be due to gross misperceptions of the attitudes and behaviors of the opposite sex. Sometimes this anxiety is severe enough to warrant referral to a psychotherapist, but in most instances a concerned, compassionate physician can be sufficiently reassuring and can make a few suggestions that help the patient develop sufficient social skills so that he (occasionally she) can learn to date with reasonable comfort.

8.1.6. Decision Making

Traditionally, the conflict over sexual decision making, particularly for girls, has been about how far to go. The fear of pregnancy, while clearly not absent in males, is more acute in females. Also, social sanctions are not evenly distributed, and girls fear the repercussions of society to a greater extent than do boys. The double standard is not yet extinct. Over the last few decades, however, a shift to relational sex has occurred and, in many instances, the decision to be made is not whether to remain virginal, but with whom to engage in intercourse. Concerns are more frequently voiced over relationships than about sexual behavior *per se*. Adolescent males have very similar conflicts, although being more genitally oriented, are apt to be more interested in the satisfaction of lust and proving one's masculinity and less in the development of a relationship. The attitudinal differences between the sexes are only relative, for many young men are also concerned about relationship issues. The wise physician can explore the choices with the young person but should be wary of offering too much advice, advice that, in any case, is not apt to be followed. In discussions about chastity or relationships, the physician should explore the advantages and disadvantages of each situation and inquire into and perhaps more sharply define the adolescent's value position and the degree of conflict with the values of parents. The intensity of peer pressure should also be explored. It is important to find out whether the young man or woman is attempting to satisfy the demands of the peer group rather than his or her own inner wishes.

8.2. Concerns of Parents

Among the most frequent parental concerns are those that have to do with the normalcy of sexual development. Once satisfied about the development of their youngsters, parents become concerned about sexual behaviors. These usually include worries about sexual orientation, over whether the adolescent is homosexual or sufficiently heterosexual. Parents are also concerned about virginity, contraception, pregnancy, venereal disease, the effects of drugs, and even sterilization for handicapped adolescents.

8.2.1. Drugs

Parents, more often than adolescents, are worried about the possible ill effects of drugs on sexual functioning. At times, teenagers themselves express such concerns. The major drugs to be considered are alcohol and marijuana (although it is clear that if the teenager has a serious addiction with heroin or amphetamines, the ill effects of those drugs are so prominent that concern over its effect on sexual functioning is relatively insignificant). While, in general, personality variables will affect the sexual behavior of adolescents addicted to these drugs more than the pharmacologic effect of drugs themselves, it may be said that heroin generally dampens sexual desire and performance, presumably because of an active endocrine effect, especially on hypothalamic mechanisms. Amphetamines may cause a temporary increase in libido and an increase in "promiscuity" as self-restraints are lowered. One must be careful not to attribute these behaviors entirely to the drug alone, since the group situation in which drugs are taken may be more significant.

The major effect of alcohol is that of disinhibition in smaller quantities and depression in larger ones. Parents are generally concerned about the possibility of sexual acting out under the influence of alcohol. The same is equally true of marijuana, but again as with alcohol, one has to differentiate between the short- and long-term effects. The long-term effects of marijuana are to decrease sexual desire and sexual performance. Daily use decreases interest in sex and the capacity to attain erections in males and the capacity for arousal and orgasm in females. On the other hand, intermittent use seems to enhance the enjoyment of sex for many. There is a vast discrepancy, however, between self-perception and actual behavior. Where marijuana does increase sexual gratification, it seems to occur because of an increased sense of pleasure from touching, a greater degree of relaxation, and from being more in tune with one's partner. For example, most people, including teenagers, report that if their sexual partner was not "high" at the same time that they were, the effect was unpleasant.

In general, parents can be reassured that neither alcohol nor marijuana will have deleterious effects on their children's sex lives provided that the teenager does not become chronically addicted to these substances.

8.2.2. Homosexuality

Parental concerns about the possible homosexual inclinations of their teenage children will be discussed under Section 8.4.

8.3. Sexual Dysfunctions

In keeping with the nomenclature developed by the Diagnostic and Statistical Manual No. 3 of the American Psychiatric Association (DSM 3), sexual dysfunctions are categorized into inhibitions of sexual desire, inhibitions of sexual excitement (impotence in the male and inhibited arousal in the female), premature ejaculation, and inhibitions of orgasm (retarded ejaculation in the male and preorgasmia in the female). These dysfunctions during adolescence have hardly been studied. Two other diagnoses listed in DSM 3, dyspareunia (pain associated with intercourse) and vaginismus (involuntary constriction of the outer third of the vagina when vaginal penetration is attempted), come to the attention of the gynecologist and usually receive appropriate evaluation and treatment. The bulk of the sexual dysfunctions associated with adult sexual life rarely are brought to the attention of the physician and even more rarely receive any systematic evaluation or treatment.

Premature ejaculation is almost ubiquitous among male teenagers, but this is rarely a complaint. Generally, the teenager is too embarrassed to even mention this concern to his physician. He may also feel that it is a matter of inexperience and that it will pass in time. That appraisal is generally accurate since approximately 90% of men eventually learn some measure of control over the ejaculatory process. Unfortunately, no studies aimed at predicting whether the overcoming of the dysfunction or the maintenance of it will take place have been attempted. This would require a longitudinal study, difficult to initiate since these young men, for the most part otherwise normal, do not enter the health care system. Impotence in adolescence is also rarely brought to the attention of the physician. Cases of primary impotence do occur, but only frequently is therapy sought for this condition. It is more likely that men with primary impotence will seek help when they are in their early twenties. Primary retarded ejaculation is seen by specialists in the treatment of sexual dysfunctions. When the physician is called on to diagnose inhibitions of sexual desire in teenagers of both sexes, he should be careful to explore for a variety of possible organic factors. In addition to systemic illness, possible hormonal deficiencies should be investigated.

Other than inhibited sexual desire, dyspareunia and vaginismus, problems related to inhibited excitement and orgasm in the female are rarely treated. Generally, the feeling is that these may represent inexperience and lack of knowledge, anxiety, and psychological immaturity that will be outgrown in the course of time. Indeed, many of these problems do resolve themselves more or less spontaneously. Our current inability to differentiate those that will be corrected by growth experience and those that will continue into adult life is a liability. Research in this area should have a high priority.

8.4. Homosexuality

The origins of homosexuality—preferential or exclusive sexual activity with a member of the same sex—are not known. The lack of specificity of various life experiences conjectured to be responsible for homosexual development, including various combinations of family relationships, have led to the speculation that a still unknown variation of the normal fetal brain programming by fetal testicular androgens may eventually be implicated.[14] For the time being, this is speculation and nothing more.

The physician commonly confronts two sorts of problems. One is the adolescent who is confused about his sexual orientation and feels threatened and isolated. The second situation involves the parents who seek the physician's help because they either know or suspect that their adolescent child is preferentially attracted to members of his own sex.

Some basic knowledge is essential in trying to understand the typical clinical situation involving sexual orientation. More than a third of adolescent boys have some homoerotic experience leading to orgasm. A smaller, but still considerable, proportion of females have had some homosexual experience during adolescence, but since less than 10% of males and probably 5% of females are predominantly homosexual, it means that the vast majority of adolescents who engage in homosexual activity develop a primarily heterosexual orientation. Caution is absolutely essential in labeling adolescent sexual activity. It would be a mistake to prematurely label such behavior as homosexual. Evidence for its preferential aspects should be sought by noting the adolescent's sexual behavior as well as his thoughts and feelings over a number of years. Adolescent boys and girls often remain uncertain of their sexual preferences for several years, so the physician has to adopt a neutral wait-and-see attitude. Too quick an acceptance of the patient's homosexuality with facile reassurance may increase the anxiety of the young person who may be far from certain of his sexual preferences. On the other hand, a judgmental censorious attitude will obviously interfere with establishing rapport. The physician's task is to help the adolescent patient reach some sort of closure about which direction he or she wishes to take (and this may take considerable time), and then to help with the social skills necessary to achieve a degree of intimacy, either heterosexual or homosexual.

Additional cues as to whether the orientation is likely to last into adult life come from the history of cross-gender activity. Among boys, cues include a history of persistent preference for girl's toys and activities, female playmates, and the company of female adults; a relative lack of interest in body contacts, sport, and other rough play with boys and fear of physical injury; occasional or frequent cross-dressing, the expressed wish to be a girl, and feminine identification and role playing. Among girls, signs would include a preference for boy's toys, activities, and clothes and a lack of interest in dolls and domestic activities; a preference for male playmates as well as for aggressive active play in sports; and cross-gender wishes. A history of cross-gender interest and activities is helpful in labeling the adolescent's behavior. On the other hand, caution is necessary since only one-third of preferential homosexual men and women report a childhood history of cross-sex behavior and preference, whereas 16% of heterosexual women report childhood tomboyism.[15]

The physician may be called upon to help the homosexual adolescent deal realistically with parents and possibly with school authorities. This is often a delicate matter involving ethical decisions, and tact, sensitivity, and good judgment are essential.

Fewer than 15% of men and 5% of women who become preferentially homosexual wish to change to a heterosexual life-style, but they should be afforded that opportunity if their motivation for change is strong. It is as wrong to try to force change on a patient as it is to withhold the possibility if the patient is motivated to undergo treatment; for this the patient should be referred to a skilled psychotherapist. Even with the help of a competent, skilled psychotherapist, there is generally about one chance in three that the patient will become successfully heterosexual.

8.5. STDs

The increase in coital activity among adolescent females, previously noted, has led to a striking increase in STDs. Just as the change in sexual activity has affected females more than males, there is also a similar change in the gender incidence of STDs. About half of all cases of gonorrhea occur in women under the age of 20, and gonorrhea is one of this country's most prevalent infectious diseases. More than 1 million cases of gonorrhea are *reported* annually in the United States, but about the same number of cases are estimated to go unreported.[16] Women bear most of the burden of this infection, for over 90% of the complications occur to them or to their offspring. Making the infection even more insidious is the "silent" nature of the disease; gonorrhea tends to be asymptomatic in 50–60% of females. (It is well to remember that the earliest symptom of gonorrhea is dysuria and, when it appears in a young female, is often dismissed as a symptom of cystitis.)

In addition to the change in sexual mores, other social factors, such as increased alcohol consumption and substance abuse and greater personal freedom, are implicated in the rise of gonorrhea. Unlike the condom, the current greater reliance on oral contraception provides no barrier to infection.

Investigations now disclose that 17% of men suffering from gonorrhea are also asymptomatic. In addition to the usual genital symptomatology, there has been an increase of pharyngeal and rectal gonorrhea, as well as reports of an increased incidence of salpingitis and gonococcal arthritis.[17] Penicillin-resistant gonoccocci create treatment problems. Those patients infected with a penicillinase-producing strain should be given a single intramuscular injection of 2 g spectinomycin hydrochloride. The treatment is always bacteriologic, and follow-up cultures for tests and cure should be obtained 3–7 and 21 days after completion of therapy. In addition to the usual treatment with penicillin, 0.5 g tetracycline hydrochloride, orally four times a day for 5 days, is also effective.

While syphilis is only $\frac{1}{40}$ as prevalent as gonorrhea, it remains a dangerous illness. More than 20,000 cases of primary and secondary syphilis were reported in the United States in 1978. More than 50% of males with syphilis are homosexual; stated another way, the homosexual male has a five times greater chance of acquiring syphilis than a heterosexual male. Despite the fact that less than 10% of all males are preferentially homosexual, the possibility of syphilis in a homosexual adolescent male should not be ignored. The adolescent homosexual should have a routine screening for syphilis and pharyngeal and rectal swabs should be obtained for cultures of gonococci.[18]

In addition to gonorrhea and syphilis, other STDs are: nongonococcal urethritis, half of which is due to *Chlamydia trachomatis*; trichomoniasis, much higher in prostitutes than in private patients; genital herpes, very prevalent among adolescents and young adults; vulvovaginal candidiasis; *Corynebacterium vaginale* vaginitis, or *Hemophilus vaginalis* vaginitis; pediculosis pubis; scabies; genital warts; lymphogranuloma venereum; granuloma inguinale; hepatitis B (common among homosexuals and prostitutes); chancroid. It is important to endeavor to trace and, if necessary, to treat the sexual contacts of patients with STDs.

8.6. Rape

Rape is defined as nonconsenting sexual intercourse. The criminal codes of most states also include statutory rape, but this is rarely enforced. Nonconsenting oral–genital and anal–genital sexual assaults are usually called sodomy, or involuntary deviate sexual intercourse, but not rape. The vast majority of rape victims are women from the ages of 18 to 45; the next most common group of rape victims are adolescent and prepubertal girls.[19]

The rape trauma syndrome includes a phase of acute disorganization followed by a long-term reorganization process. The acute phase is characterized by disturbances of sleeping and eating, excessive fears, repetitive thoughts about the assault, and a variety of physical complaints. The family's emotional response may create additional disruptions.

Most frequently the family has the need to blame someone—the rapist, the child, or themselves. Most psychological damage is caused if the child is blamed, but damage is often created by the parents blaming the assailant or themselves. Murderous rage directed toward the rapist may remind the child of the violence of the assault itself; it also may be frightening reminder that her own family is capable of extreme aggression. When parents blame themselves, it may increase the child's distrust of people or may lead to overprotective behavior by the parents.[20]

In adolescence, the long-term reorganization phase may be manifested by nightmares, continued sleeping disturbances, and the emergence of phobias. School problems are frequent, including poor academic performance, peer relationships, truancy, and school phobia. Sexual fears and concerns are particularly important, for the adolescent girl is often not very knowledgeable about sexuality, and false beliefs and attitudes about sexual functioning may develop.

The management of rape in the young adolescent requires management of parental fears and other emotional reactions. If the parent is relatively calm, the presence of a stable, comforting parent is an asset during the interview and examination. If the parent is extremely anxious, the parent becomes a liability. A discussion with the parents is essential. Parental anxiety must be decreased; if not, parental anxiety will increase the anxiety level of the young adolescent. Discussion should include education about appropriate parental behaviors and what adolescent behaviors may be anticipated.

If the young adolescent girl is being examined in the hospital room, a family member may be of great help in reducing the anxiety-provoking strangeness of the setting. Examination of external genitalia may reveal lacerations, bleeding, or other physical trauma. Surgical competence in repair techniques may be required.

The physician should try to avoid making a judgment as to whether the victim is telling the truth. He need only be an objective observer and a recorder of data. In most major cities, a support person trained by a Rape Crisis Center is generally on call at the Emergency Room. This person can provide both crisis intervention and the option of continued support in ensuing days and weeks. Most emergency rooms also have ''rape kits'' for the collection of specimens, which simplify and expedite the physician's job.

8.7. Incest

It is uncertain just how frequently incest occurs. It has been estimated that first-cousin incest, the most frequent of all types, involves perhaps 3–4% of the United States population. Less frequent is incest with siblings, then other members of the family, and least of all with fathers and mothers. While, for almost all people, the seduction of a child or adolescent by an adult is morally repugnant because of actual or implied force or coercion, there is less certainty about the long-term effects. Certainly incest psychologically damages many people; for example, the guilt over sex with a family member may later on interfere with sexual desire or performance. If the adult tells the adolescent that the activity has to be kept secret, even if the sexual activity is pleasurable, it becomes associated with sinfulness. Sex, love, pleasure, and punishment become mixed and blurred. The situation in which sexual activity with family members takes place may cause great anxiety and tension. The fear of retaliation for revealing the secret may be severe.[21] Incestuous activity is often legally defined as regularly occurring intercourse between family members, but there may be noncoital heterosexual or homosexual sex play that may be upsetting or painful.

What is not known is how often this type of activity leads to permanent damage and, in contrast, how often people apparently cope with it without any long-term consequences. Even if it does not later result in specific sexual problems, the blurring of generational boundaries may interfere with a child's or adolescent's development. Problems with authority or unsatisfactory adult relationships may be a consequence. Since incest generally occurs in families that are disorganized without good role definition and occurs often in the presence of alcoholism, drug abuse, and mental illness, long-term damaging effects may be the consequence of the total family situation rather than of incest itself.

When incest involving a teenager is reported to the physician, the first concern is possible pregnancy, physical damage, or disease. If a pregnancy occurs and abortion is rejected as an option, the role confusion that arises if the baby's father is also its grandfather or its mother also is its sister causes extraordinary complications in family relationships. In those circumstances, if it is acceptable to the family, adoption is often a wise choice.

If the adolescent denies the report of incest, the physician should refrain from putting himself in the position of judging who is telling the truth. Often, in such instances, a local social service agency may provide for long-term care, for it is clear that if there is even an accusation of incest, the family needs help.

Health professionals must guard against venting moral outrage. The family, including the teenager, needs understanding, support, and acceptance. This does not imply that the physician should condone incest. The physician should move slowly, for precipitous arrest of the incestuous adult may seriously damage or even destroy the family. Sometimes the removal of the child is preferable to jailing the father. The dynamics of the family must be understood and adequate treatment directed to the whole family unit.

9. UNCOMMON SEXUAL DISORDERS

Sexual disorders that appear relatively infrequently include genetic defects that lead to ambiguous external genitals (intersexuality); types of gender dysphorias, most notably transsexualism; and paraphilias, such as transvestism and fetishism.

9.1. Intersexuality[22,23]

9.1.1. Klinefelter Syndrome

Klinefelter syndrome occurs in approximately 1 in 500 live male births, making it one of the most common sex chromosome disorders. In this disorder, one finds the presence of an extra X chromosome, typically 47XXY. Since external anatomy is completely normal in a newborn boy with Klinefelter syndrome, the condition may not be diagnosed before late adolescence or even early adulthood. Typically, on examination, one finds small testes, somewhat smaller than usual penis size, and persistent gynecomastia. It is only the gynecomastia that usually comes to medical attention in early adolescence. In most cases, there is a low level of testosterone leading to delayed sexual development in adolescence. By late adolescence or early adulthood, one finds low sexual desire and sometimes impotence. Those boys with body changes in puberty or in early adolescence are subject to disturbances in body image and gender identity.

Although many adolescents with Klinefelter syndrome turn out to be heterosexual, one finds a higher than normal prevalence of homosexuality, bisexuality, transvestism, and serious gender dysphoria. Infertility is almost invariable and is irreversible, but the gynecomastia is amenable to plastic surgery. On occasion, one finds dramatic reversal of low libido and even of impotence with testosterone replacement therapy, but not all respond to such treatment. When the diagnosis is made in adolescence, supportive therapy should be instituted, directed at preventing or alleviating sexual identity problems.

9.1.2. Turner Syndrome

Turner syndrome, which as a 45XO chromosomal configuration, occurs in approximately 1 in 2500 live female births. The most notable features are short stature and amenorrhea. Generally, development is normal until puberty. Instead of fully formed gonads, the single X chromosome disorder leads to the development of primitive streak gonads. These produce few estrogens and other sex hormones. The absence of sex hormones leads to absent menses, lack of breast development, and poor skeletal growth. The physical differences between girls with Turner syndrome and peers causes problems in self-esteem. Hormone replacement therapy can bring about menarche.

The physician should concentrate on two facets of Turner syndrome, the need to manage the replacement steroids and counseling including increasing social skills. These, in turn, are dependent upon modifying the poor self-esteem of these adolescent girls. The clinician should always be truthful in explaining the nature and consequence of the chromosomal anomaly. Deception will interfere with the development of trust so essential to appropriate counseling. Discussions, of course, should be tactful and sensitive. While infertility is highly probable, it will help to explain to the patient that she can become a mother by adoption.

Hormone replacement therapy should begin at about the time when puberty would normally occur. Of course, the use of estrogens to induce menarche and the secondary sex characteristics essential to the girl's normal sexual identity also carries with it the cost of closing off the growth centers of the long bones, further curtailing growth.

9.1.3. The Adrenogenital Syndrome

The excessive production of adrenal hormones due to hypersecretion of adrenocorticotrophic hormone and, in turn the consequence of blockage in the synthesis of cortisol leads to excessive masculinization of females. Generally, these cases are diagnosed in childhood and treatment begun early in life. There is a tendency on the part of these girls, even when treated early, to have difficulty forming close heterosexual relationships. They have a greater frequency of bisexual fantasy and experience and show some inhibition of erotic interest. Some of these consequences may be due to the chronic nature of treatment, including genital surgery, continuing use of medications, and repeated pelvic examinations.

9.1.4. Testicular Feminization Syndrome

This is a rare disorder in which in a genetic male 46XY fetus the development of male external genitals fails to occur because of tissue insensitivity to the testosterone that is normally produced during fetal life.

Ordinarily, even when the diagnosis is made at birth, these children are raised as girls. These are genetic males with testes but with the external appearance of girls, and they can be raised as girls in a remarkably normal fashion. In adolescence, they can be very feminine-looking and are generally feminine in a broad range of behaviors and attitudes.

A special syndrome with similarities to testicular feminization has been found among Dominican male pseudohermaphrodites[24] with decreased dihydrotestosterone production due to a 5-α-reductase deficiency.[25] Unlike males with androgen insensitivity, all but 1 of 18 Dominican subjects studied changed to a male gender identity during or after puberty. This condition appears to be different from the usual form of testicular feminization that is due to a biochemical defect involving androgen binding to receptor sites in target tissues.

9.1.5. Vaginal Agenesis

The failure to develop a vagina may be a simple consequence of incomplete müllerian duct development in the presence of normal ovaries, tubes, and uterus. In other instances, there is faulty development of internal sex organs as well. These patients have been reared as girls and generally have a normal female core gender identity, heterosexual orientation, and feminine sex role preferences. Vaginoplasty is usually safe and successful and permits full sexual function, including orgasm.

9.2. Gender Dysphorias

Gender dysphoria as a consequence of intersexuality has been partially discussed above. It is important that a physician recognize that anomalous genitalia or atypical secondary sex characteristics can cause not only sexual identity problems but even psychotic depression and suicide. That is why, as Gadpaille suggested, "in cases of true hermaphroditism, all organs of the sex opposite to the sex of rearing should be surgically extirpated so that the physician can honestly assure the adolescent that no contradictory sex organs exist."[15]

9.2.1. Transsexualism

The gender dysphoria associated with primary transsexualism becomes particularly intense in early adolescence. The changes to an adult body and the development of adult genitalia intensify the feeling the individual has of being trapped in the wrong body. It is not uncommon for some early adolescents who feel this way to surreptitiously obtain and use estrogens to change their anatomy. It is rare for early adolescents to press for sex reassignment surgery, and it is unheard of for a youth to have this procedure performed. There are only three reports of a nonsurgical successful shift of core gender identity in an adolescent or older transsexual.

9.2.2. Transvestism

Although the Diagnostic and Statistical Manual III of the American Psychiatric Association lists transvestism as a paraphilia instead of a gender dysphoria, there is an overlap between transvestism and transsexualism, creating some diagnostic confusion. Not all patients with transvestism can be sharply demarcated from those with a transsexual disorder. Some adult transvestites have been cross-dressing since they were young children. More develop the overt syndrome as they approach adulthood. Transvestism also has an overlap with fetishism, a paraphilia. Adolescent boys not infrequently steal the underclothes of mother or a neighbor, and this may be a reason why a physician is consulted. These garments enhance erotic arousal and are generally used in masturbation. They may remain as fetishistic objects for erotic arousal or may then be included in a more extensive attempt to dress in female garments as part of a greater identification with the female.

9.3. Paraphilias

Transvestism and fetishism have been briefly discussed under Section 9.2. Most paraphilias are not fully developed until adult life. Some voyeuristic and exhibitionistic acts are so common in adolescence that it is difficult to make a definite diagnosis unless it is a very persistent and compulsive pattern. If the disorder is fully developed, and particularly if the disorder brings the adolescent into trouble with the law, therapy should be instituted. Certainly this is true with the more serious paraphilias, such as pedophilia or sadomasochism. In most instances of paraphilia, the physician should refer the patient to a psychiatrist.

The term *paraphilia* is a word derived from the Greek root "para" meaning alongside of and "philia" meaning love. It was first used by the pioneer American psychiatrist, Adolph Meyer, and was resurrected by the DSM-III Committee on Psychosexual Disorders in order to somewhat neutralize the stigma of deviant sexual behavior. This allows a physician to help a distressed patient without the repugnance, or even moral outrage, that might inevitably accompany more pejorative labels. The range of paraphiliac behavior is great. At one end of the spectrum is the fetishist who is carrying out a victimless act. At the other end is the lust murderer whose aberrative script requires extreme violence to accomplish sexual arousal. The primary care physician has the task of retaining his equanimity and detached concern long enough to obtain a reasonably accurate history, enabling him to make an appropriate referral to a psychiatrist. In his contact with the patient, the physician should emphasize the inherent dangers of the behavior, both legal and social, and stress the importance of psychiatric care. The physician should be considerate and compassionate while indicating willingness to continue medical care during and after psychiatric treatment.

REFERENCES

1. Lief HI: Sexuality and sexual health, in Lief, HI (ed): *Sexual Problems in Medical Practice*. Monroe, Wisc, AMA Press, 1981, pp 9–16
2. Schwartz MF, Money J, Robinson K: Biosocial perspectives on the development of the proceptive, acceptive and conceptive phases of eroticism. *J Sex Marital Ther* 7:243–255, 1981
3. Zelnik M, Kanter JF: Sexual and contraceptic experience of young unmarried women in the U.S., 1976 and 1971. *Fam Plann Perspec* 9:55–71, March/April 1977
4. Furstenberg FF, Lincoln R, Menken J,: *Teenage Sexuality, Pregnancy and Childbearing*. Philadelphia, Univ of Pennsylvania Press, 1981
5. Reiss IL, Furstenberg FF: Sociology and human sexuality, in Lief, HI (ed): *Sexual Problems in Medical Practice*, Monroe, Wisc, AMA Press, 1981, pp 53–70
6. Delamater J, MacCorquodal P: *Premarital Sexuality*. Madison, Univ of Wisconsin Press, 1979
7. Reiss L: *The Social Context of Premarital Sexual Permissiveness*. New York, Holt, Rinehart & Winston, 1967
8. Persky H, Lief HI, Strauss D, et al: Plasma testosterone level and sexual behavior of couples. *Arch Sexual Behav* 7:157–173, 1978
9. Raboch J, Bartak V: Menarche and orgastic capacity. *Arch Sexual Behav* 10:379–382, 1981
10. Lief HI, Berman EM: Sexual interviewing of the individual patient through the life cycle, in Oaks, WW, Melchiode, GA, Ficher, I (eds): *Sex and the Life Cycle*. New York, Grune & Stratton, 1976, p 1
11. Group for the Advancement of Psychiatry: *Assessment of Sexual Function: A Guide to Interviewing*, GAP Report No. 88, Committee on Medical Education. New York, GAP, 1973
12. Litt IF, Cohen MI: Adolescent sexuality. *Adv Pediatr* 26:119–136, 1979
13. American Psychiatric Association: *Diagnostic and Statistical Manual of Mental Disorders*, ed 3., Washington, DC, Am Psychiatric Assoc, 1980
14. Bell AP, Weinberg MS, Hammersmith SK: *Sexual Preference*. Bloomington, Indiana Univ Press, 1981
15. Gadpaille WJ: Sexual identity problems in children and adolescents, in Lief, HI (ed): *Sexual Problems in Medical Practice*. Monroe, Wisc, AMA Press, 1981
16. Diamond M, Karlen A: Sexually transmitted diseases, in Lief, HI (ed): *Sexual Problems in Medical Practice*. Monroe, Wisc, AMA Press, 1981
17. Donald WH: The changing pattern of sexually transmitted diseases in adolescents. *Practitioner* 222:383–385, 1979
18. Jaffe LR, Morgenthan JE: Syphilis and homosexuality in adolescents. *J Pediatr* 95(6):1062–1064, 1979
19. Burgess AW, Groth, AN, Holmstrom, LL, et al: *Sexual Assault of Children and Adolescents*. Lexington, Mass, Lexington Books, 1978
20. Miller WR, Fordney DS: Sexual exploitation and aggression, in Lief, HI (ed): *Sexual Problems in Medical Practice*. Monroe, Wisc, AMA Press, 1981
21. Renshaw DC: *Incest—Understanding and Treatment*. Boston, Little, Brown, 1982.
22. Kolodny RC, Masters WH, Johnson VE: *Textbook of Sexual Medicine*. Boston, Little, Brown, 1979
23. Money J, Ehrhardt AA: *Man and Woman, Boy and Girl*. Baltimore, Johns Hopkins Press, 1972
24. Imperato-McGinley J, Peterson RE: Male pseudohermaphroditism: The complexities of male phenotypic development. *Am J Med* 61:251–272, 1976
25. Imperato-McGinley J, Peterson RE, Gautier T, et al: Androgens and the evolution of

male gender identity among male pseudohermaphrodites with five-alpha-reductase deficiency. *N Engl J Med* 300:1233–1237, 1979

RECOMMENDED READING FOR TEENAGERS AND ADULTS

The following list is selected from the SIECUS (Sex Information and Education Council of the U.S.) guide to "Family Reading about Sexuality."

Recommended Reading for Early Teens

Gordon S: *Facts about Sex for Today's Youth*, ed revised. Syracuse, New York, Ed-U Press, 1979 (A short, direct approach in explaining anatomy, reproduction, love, and sex problems. Includes slang terms when giving definitions, and a section answering the ten most common questions children ask. Well illustrated.)

Gordon S: *Facts about V.D. for Today's Youth*, ed revised. Syracuse, New York, Ed-U Press, 1979 (Up-to-date, accurate information written in clear and simple language. Stresses prevention.)

Hamilton E: *Sex with Love: A Guide for Young People*. Boston, Beacon Press, 1978 (Includes discussion of the rituals of early dating and fulfilling the body's need for affection and sexual expression.)

Johnson EW: *Love and Sex in Plain Language*, ed revised. New York, Harper & Row, 1977; New York, Bantam Books, 1979 (Provides basic information on sexuality and emphasizes that sexuality should always be seen in the context of one's total personality and expressed in responsible, respectful interpersonal relationships.)

Johnson EW: *Sex: Telling it Straight*, ed new. New York, Harper & Row, 1979 (A simple but honest treatment of those topics in human sexuality of greatest concern to adolescents. Written for teenage slow readers, especially those within a ghetto environment and presents positive views on sex without preaching or moralizing.)

Johnson EW: *V.D.—And What you Should Do about It*. ed new revised. New York, Harper & Row, 1978 (For any grade. Facts in context. No horror pictures.)

Kelly GF: *Learning about Sex: A Contemporary Guide for Young Adults*, Woodbury, New York, Barron's Educational Series, Inc., 1977 (Without neglecting basic factual information, focuses on attitudes and the process of sexual decision making.)

Pomeroy WB: *Boys and Sex. Girls and Sex*, ed revised. New York, Delacorte Press, 1981 (These classic sexual guides for teenage boys and girls have been updated.)

Recommended Reading for Later Teens

Bell R, Bowie S, Evans R, et al: *Changing Bodies, Changing Lives*. New York, Random House, 1980 (A forthright, nonjudgmental book by the authors of *Our Bodies, Ourselves* for teens which confronts their real concerns about sex and relationships. Highly recommended.)

Comfort A, Comfort J: *The Facts of Love: Living, Loving and Growing Up*. New York,

Crown, 1979 (A dynamic book about sexuality, ideal as a catalyst for conversations with young people.)

Demarest RJ, Sciarra JJ: *Conception, Birth and Contraception: A Visual Presentation*, ed 2. New York, McGraw–Hill, 1976 (A fine, concise pictorial presentation of human reproduction. Text is simply stated, expanding upon the illustrations.)

Gordon S, Wollin M: *Parenting: A Guide for Young People*. New York, Sadlier, 1975 (A thoroughly modern exposition to prepare potential parents for mature parenting roles.)

Hanckel F, Cunningham, J: *A Way of Love, a Way of Life: A Young Person's Introduction to What It Means to Be Gay*. New York, Lothrop, Lee & Shepard Books (William Morrow), 1979 (A unique, sensitive book written by people who are having the experience for people who want to understand it.)

Lieberman EJ, Peck E: *Sex and Birth Control: A Guide for the Young*, ed revised. New York, Harper & Row, 1981 (Written to encourage sensible and responsible use of birth control and to encourage young people to develop principles and values by which they will live their sexual lives.)

Mazur RM: *Commonsense Sex*. New York, Beacon Press, 1973 (Based on the premise that sex is a positive aspect of human personality and concludes with a suggestion of a liberal religious framework for decision making.)

Morrison E, Starks K, Hyndman C, Ronzio N: *Growing up Sexual*. New York, Van Nostrand, 1980 (Unique view of patterns of human sexual development based on anonymous autobiographical papers by students in a college human sexuality course.)

Recommended Reading for Parents

Adams J: *Sex and the Single Parent*. New York, Coward, McCann & Geoghegan, 1978 (Realistic portrait of intellectual and emotional struggles. Recommended for practitioners dealing with single-parent families, as well as for the members of single-parent families themselves, including adolescents.)

Arnstein HS: *The Roots of Love: Helping Your Child Learn to Love in the First Three Years of Life*. Indianapolis, Bobbs–Merrill, 1975 (Shows how physical, mental, and motor development in the young child relate to concurrent emotional development. Suggests methods for parents to use in helping their infants' growing ability to love and form human attachments.)

Bernstein AC: *The Flight of the Stork*. New York, Delacorte Press, 1978 (A book for parents, using theories from Piaget and Kohlberg, illustrating how to explain reproduction in terms of children's understanding at various age levels.)

Calderone MS, Johnson W: *The Family Book about Sexuality*. New York, Harper & Row, 1981; New York, Bantam, 1982 (A creative, comprehensive approach to a family's understanding of the sexuality and sexual concerns of all its members. Includes encyclopedic glossary and terms.)

Corsaro M, Korzeniowsky C: *STD—Sexually Transmitted Diseases: A Commonsense Guide*. New York, St. Martin's Press, 1980 (Well-organized, clearly written, and up-to-date guide of the most common STDs. Stresses alerting for prevention and need for prompt attention to symptoms.)

Fairchild B, Hayward N: *Now That You Know: What Every Parent Should Know about Homosexuality*. New York, Harcourt Brace Jovanovich, 1979 (Informative, sensitively written guide for parents of homosexuals. Highly recommended.)

Greenberg S: *Right From The Start*. Boston, Houghton Mifflin, 1979 (Recommended for parents who want to help children of either sex build on their own particular strengths. Redefines motherhood, fatherhood, and family power relationships and demonstrates how differential treatment of boys and girls hinders their development.)

Oettinger KB, Monney E: *Not My Daughter: Facing Up to Adolescent Pregnancy*. Englewood Cliffs, NJ, Prentice–Hall, 1979 (Helpful for parents and for those seeking perspectives on the problem in their search for preventive measures. Stresses need for early communication between adults and teens.)

Pogrebin LC: *Growing Up Free: Raising Your Child in the '80s*. New York, Bantam Books, 1981 (Covers child-rearing from conception to majority. Emphasizes nonsexist sex education, parity parenting, and gender-neutral attitudes. Highly recommended.)

Pomeroy W: *Your Child and Sex: A Guide for Parents*. New York, Delacorte Press, 1976 (Gives parents a better understanding of their own sexuality, both in their marriage relationship and in their relationships with their children. Also deals with ways and means of talking about sex to children at various age levels, from the very young to post adolescents.)

Silverstein C: *A Family Matter: A Parents' Guide to Homosexuality*. New York, McGraw–Hill, 1977 (Written for parents with a homosexual child, examining the realities of the situation and suggesting how to turn the experience into a positive relationship.)

Uslander AS, Weiss C, Telman J: *Sex Education for Today's Child: A Guide for Modern Parents*. Chicago, Association Press/Follett, 1977 (Provides model answers to typical questions asked by children concerning sex.)

CHAPTER 5

Teenage Pregnancy

Helen O. Dickens and Dale M. Allison

1. INTRODUCTION

1.1. Incidence of Pregnancy

The increased incidence of teenage pregnancies can be viewed as a result of demographical, biological, and social factors. Throughout human history, there have always been people who began childbearing and child-rearing at very early ages. At times and particularly in some cultures, becoming a mother or father in the early teens has been accepted and even encouraged. In other societies, such behavior has been frowned upon and the young people punished or, at best, neglected. Today, it is no longer believed that it is in the best interest of our communities or in the best interest of the young families themselves to neglect or ignore them. They are an integral part of our society and should be helped to become contributing citizens and to develop a strong, stable family life whenever possible.

Adolescents in the United States have childbearing rates that are among the world's highest. About 10% of American teenagers get pregnant and 6% give birth each year. Among the world's industrial countries, only Rumania, Hungary, Yugoslavia, Bulgaria, East Germany, and Czechoslovakia have higher teenage fertility rates. Adolescent fertility in the United States is higher than in many less developed countries, i.e., the Philippines, Tunisia, and East Malaysia. The much-talked-about postwar baby boom that increased the number of teenagers in the general population also increased the number of pregnant teenagers. The culmination of the years of improved nutrition, the drop in the age of men-

Helen O. Dickens • Department of Obstetrics and Gynecology, School of Medicine, University of Pennsylvania, Philadelphia, Pennsylvania 19104.   Dale M. Allison • Department of Obstetrics and Gynecology, School of Medicine, University of Pennsylvania, Philadelphia, Pennsylvania 19104.

arche for the past few decades, makes the biological age at which girls can become pregnant lower; therefore, the drop in age of menarche also contributed to their increased numbers in the United States. It has been stated that teenage pregnancy begins when it becomes possible. School-age pregnancy was and is increasingly possible at younger ages.[1]

Birth to adolescents aged 15 and younger increased by 61% from 1960 to 1977. Premarital sexual activity has increased by two-thirds in the 1970s especially among 15 to 19-year-olds living in the metropolitan areas. The sexual activity of whites aged 15–17 has doubled. The distinction of race, socioeconomic status, residence, and religion can no longer be associated with the age at the first sexual intercourse. Although the age at first sexual intercourse is younger, 16.4 for white and 15.5 for black women, marriage rates for teens have dropped 4% for whites and 45% for blacks.

One-third of all teens always use a method of contraception, and one-half use a method of contraception at first intercourse. The condom is the major method of choice at first intercourse. The major reasons for nonuse are ignorance, unexpected intercourse, or pregnancy. A reported increase in sexual activity among young people is coupled with an inadequate use of contraceptives especially by the youngest teenagers (12–15 yr) who probably get pregnant during their first or second sexual relationship. New data suggest one-fifth of the adolescents who get pregnant conceive in the first month of sexual activity and one-half in the first 6 months.

Not all pregnancies end in birth. Of the 1.1 million teen pregnancies in 1978, 38% were terminated in abortions, 13% were miscarriages, 22% were born out-of-wedlock, 17% of adolescent pregnancies were premaritally conceived, and 1% were legitimate births premaritally conceived. Among 15 to 17-year-olds, pregnancies increased by 4%; however, abortions increased by 45%. For teens under 15, pregnancies increased by 5% and abortions were up by 17%.

Better use of contraception has brought the teen pregnancy rate down. The teen pregnancy rate rose from 10 to 11% between 1973 and 1978. The sharpest fall was between 15 to 17-year-olds, where pregnancy fell from 26 to 21%. The rate of teen pregnancy is currently rising at a slower rate than the rate of sexual activity.

Reducing the incidence of repeat teen pregnancies is a major focus of health care for the adolescent who chooses to carry her infant to term. Adolescents who have had one pregnancy are at risk of having a second pregnancy. Consistant use of contraception has reduced the number of repeat pregnancies in the first year. In 1978, 15% had repeat pregnancies, compared to 22% in 1971. Repeat pregnancies two years after the first pregnancy are 30%, compared to 50% in 1971. Nationally in 1976, 96% of women 15–19 who delivered a first baby out-of-wedlock kept their babies, thus resulting in 1.3 million children to be living with 1.1 million teenagers.

Death related to pregnancy and its complications is 35% higher in 15-to 19-year-old mothers and 60% higher in adolescents 14 years old and under. The death rate for infants born to teenagers under 18 is nearly twice that for infants born to women in their twenties, and the incidence of low birth weight is estimated to be twice as high in this group as among women in their twenties.

As mentioned previously, births to teenagers have increased primarily as the result of a marked increased in the number of adolescent women. Forty-one percent of unmarried teens thought they could not become pregnant. However, of those who did become pregnant, very few wanted to be pregnant. Eight in ten premarital teenage pregnancies and two-thirds of premarital teenage births are unintended. Teenagers represent only 18% of all sexually active women capable of becoming pregnant, but they account for 16% of all out-of-wedlock births and 31% of all abortions.[2]

The disciplines of pediatrics and obstetrics and gynecology have important and appropriate roles to play in adolescent pregnancy both in prevention as well as in stimulating the fair and impartial development of the nation's resources for the preadolescent, the nonsexual adolescent, and the pregnant adolescent and her offspring.

1.2. Reasons for Adolescent Pregnancy

Teenage childbearing is a serious and growing social, economic, and health problem in the most highly developed countries in the world. Often the problem is compounded because the births are out of wedlock. Numerous reasons have been cited for teenage childbearing.

The age of menarche has decreased, making it possible for a young woman to bear children at an earlier age. Our educational and economic standards have increased, causing the crisis to be traumatic to the family and costly to society. The cost to support a mother and child on welfare, medical and social services, and any other children she may have was $2250 in 1979.[1]

Many teenagers do not understand the likelihood of pregnancy, because they have partial development of their analytical probabilistic thought processes. Teenagers need the time and opportunity to connect information and behavior.[3]

The young women, who tend to be dependent, passive, inarticulate, low in self-esteem, and communicate poorly with parents or not at all, especially regarding sex, are often exposed to sex early and are at risk of pregnancy. They often lack success in school especially when they have no goals to further their education. Such young women may lack self-confidence and want more kindness and reassurance from physicians, other professionals, and volunteers. The adolescent can be led to clarify her own values and goals through self-confidence.

Teenage mothers are first and foremost teenagers and are at the maturation level of their peers. They are "risk-takers"; many know that they can become pregnant, yet some choose to take the chance. Some authorities have labeled promiscuity the "juvenile delinquency" of female adolescence. However, most girls seem to have a strong attraction for their male partners.

Many female adolescents may feel lonely and isolated. Sexual intercourse offers a means to be held and emotionally loved. This helps to affirm one's femininity.

Feelings of self-destruction, anger and aggression toward authority, lack of responsibility for one's own body and actions, and desperate pleas for attention and help have been noted in girls who become pregnant. The child mother is seven times as likely to attempt suicide on a nationwide average.

Adolescent pregnancy may occur to compete with or identify with the mother. It is not uncommon for teens to begin childbearing at about the same age as their parents began childbearing.

Some girls become pregnant because they do not foresee other opportunities in their lives and do not have high aspirations for school or careers. Pregnancy can be a way of "dropping out" of the maturation process of adolescence. Some adolescent mothers drop out of school even when the schools allow them to remain. The new mother receives attention from her family for her pregnancy and the arrival of the newborn. In her family, she is elevated in status from a "child" (or adolescent) to "child mother" and is expected to be more responsible.

Peer pressure to begin a sexual relationship and, in some groups, to have a baby can be related to early sexual activity. Girls can feel that to be popular they must engage in sexual intercourse.

The sexual environment has opened earlier exposure to sexual behavior. Early onset of dating, limited parental control, open contact with the opposite sex, and dress, advertising, movies, literature, music, and language have become apparent to all. The social message delivered is that sex is good, fun, free, and the best "high" of all—who could possibly resist?

There are those who do resist and are not pregnant because they choose not to be sexually active. Some adolescent pregnancies are the result of rape (see Chapter 8) or incest. Incest is the result of a relationship with a father, brother, grandfather, or uncle. These relationships tend to be hidden, and the teen will report that her partner is not available. Incest is not reported immediately and is hidden in the family throughout the pregnancy and may not be discovered for years, if at all.

There are still a large number of adolescents in our society who abstain, delaying sex for education, economic stability, and other goals. These adolescents also deserve better health facility support and education support. The nonpregnant sister may feel that the pregnant sister gets all of the attention and support and is at a higher risk of pregnancy than the girls in the general population.

Adolescent pregnancy is a result of early onset of menses and many sociological and psychological reasons. It is also present due to the failure of society to teach family life education and provide contraception for those who are sexually active at an early age.

The public schools in Philadelphia suppressed a school paper, *Town Crier*, of October 26, 1977, in which three articles regarding adolescent sexuality appeared under the headline "Pregnant Teens Face Painful Choices." One article told the story of a young couple who were married because of the pregnancy, how they tried to work it out, and the difficult solution. The question was asked, "Why were contraceptives not used?" and the girl could not respond. In a sense her case is identical to a million others. The second article began "Planned Parenthood Association Has Described Teenage Pregnancy as an Epidemic." The article goes on to point out that in the past when a pregnancy occurs, there were only two options: one, the girl could travel to another state, have the child, and put it up for adoption. The possibility of marriage was given as the second option. Today, young women are less likely to marry but rather raise their babies themselves. The article then went on to tell the story that we all know: dropping out of schools, failing to find a job, and ending up on welfare. The third article stated at the beginning that it in no way represented counseling but discussed the pill, intrauterine device (IUD), diaphragm, and condom. This article further give information to contact Planned Parenthood, supplying the local telephone number. The article so disturbed the principal, faculty, and board of education that the publication was suppressed. It is of further interest that these teenagers took their cases to the federal court because of suppression of their civil rights and won. This is the kind of furor that was present in the high schools of Philadelphia in 1977 regarding even the discussion by the pupils in print of the problems relative to teenage pregnancy and contraception by the students.

2. COST TO SOCIETY OF ADOLESCENT PREGNANCY

The financial cost of adolescent sexuality and pregnancy to American society is enormous. In 1976, 215 million dollars was spent for teenagers in family planning clinics for sterilization, abortion, social work, contraception, and sex education. Adolescents are three to five times more likely to depend on the government to pay for delivery. Data from Rhode Island indicate 69% of births to girls under 15 and 46% to those 15–19 are paid for by public funds. Teens are also twice as likely to have to pay for delivery out their own pocket as older women, who are more likely to have insurance coverage.

The adolescent mother pays for her pregnancy in many ways. Educational opportunities and career plans are likely to be frustrated, particularly for the younger teen. At 29 years of age, women who delayed childbearing to their early twenties were four to five times more likely to have completed college than the adolescent mothers. Low education also accounts for the fact that teenage mothers earn approximately half the income of their peers who delayed childbearing. Hence, families headed by young mothers are seven times as likely to be poor.

For the pregnant teen who marries, stability is less likely and the divorce rates are high. Forty-four percent of the women who gave birth between 14 and 17 are divorced 15 years later. She will also have 50% more births than women who are later childbearers.[2]

The adolescent father usually feels he is in love with the girl when she becomes pregnant. He views his problems as coping with being a father and concerns about his parenting skills. He knows he has financial responsibilities and is concerned about employment, restrictions on his freedom, and getting along with the unwed mother and the extended families on both sides. In some instances the girl's family is angry with him and tends to exclude him. The adolescent father senses isolation and feels he has few places or people to whom he can turn. Young men have been known to suffer psychiatric problems due to the pressures and responsibilities they feel.[4] Educationally, the adolescent father is also disadvantaged. He is two-fifths less likely to have graduated from high school than his peers. By age 29, those who were teen fathers are only half as likely to have completed college.

The grandparents of the teen mother are likely to have been early childbearers themselves. Twenty-two percent of the mothers 17 and under and 43% of the young fathers had a teen parent.[2] Mothers of the adolescent mothers often become the primary caretaker of the new infant when the teen mother delivers and returns to school. This places a burden on the grandmother who may only be in her thirties and sees the first chance of freedom from her own childbearing being removed from her.

Children who are born to adolescent mothers also pay a price. Studies show that children of teenage parents also suffer educational and cognitive defects. They tend to have lower IQ scores than children of delayed childbearing and are more likely to repeat at least one school grade. However, when grandparents or the child's father helped the teen mother care for the baby, the child's cognitive development was found to be superior to that of children brought up by the young mother alone.[2]

3. HEALTH PROBLEMS AND TEEN PREGNANCY

3.1. Compliance

Ambivolence about a pregnancy is not uncommon, particularly when most adolescent pregnancies are unplanned. Therefore, having to conform to physical examination and somewhat uncomfortable laboratory procedures can cause avoidance behavior.

Complete and open explanations given in a direct and personable manner helps reduce anxiety. The physician should be clear and honest with explanations. Correct terminology, followed by a simplified explanation, aids in patient understanding and willingness to comply. Adolescents are sensitive and respond negatively if they feel that they are being spoken to as a child; likewise, if the explanation is too technical, the adolescent may feel inhibited and not ask appropriate questions. Eye-to-eye contact helps patients to feel their problem is important and the care provider is interested. Questions should be encouraged directly on a one-to-one basis or by having the teen write down questions and bring them to each visit. If the adolescent feels respected by the health team, compliance is more readily received.

3.2. Pelvic Examination and the Pregnant Teenager

Most pregnant adolescents, especially 15 years of age and younger, have never had a pelvic examination. If they have had a negative experience previously, the advent of the pregnancy makes them more apprehensive to the examination. It is well to take a detailed history while establishing as relaxed an atmosphere as possible, establish trust, and let her know that you are her advocate. This may be time-consuming, therefore, ample time must be allowed.

Following the history, the teen should be sent into the examining room with an attendant (professional or family) to undress and put on gown or robe. Upon entering the examination room, the examining person again tries to put the adolescent at ease explaining each step before it is taken. Sometimes it is advisable to have her spread her own genitalia, warm the speculum prior to use, and handle and/or insert the speculum herself. If the resistance is great, a second or third appointment should be considered to complete the examination. In no case should the patient be forced, held down physically, or verbally chastized. In extreme cases, pelvic examination can be deferred until later in the pregnancy. An occasional patient may need an epidural during labor, and emotional support before a successful pelvic examination is possible. Psychiatric consultation has been found to be helpful if only to advise against pelvic examination at all until advisable. This is especially important in cases where there has been rape or incest.

Adolescent pregnancies tend to be normal pregnancies with low hemoglobin and hematocrit levels noted throughout pregnancy and an increased rate of preeclampsia at term.

3.3. Iron and Folacin Intake

Adolescents have greater nutritional requirements in relation to their body size than adult women. The additional nutrient demands of pregnancy and poor dietary habits may compromise their growth potential and increase their risk during pregnancy. Reports of nutritional surveys indicate that adolescence is a period of increased risk for iron deficiency. Black pregnant women have been reported to have lower hematocrits than pregnant white women. This difference, which was income corrected, suggests that race-specific norms would refute the number of black considered anemic. Nevertheless the presence of sickle cell disease should be ruled out by routine testing (i.e., sickle cell preparation).

It has been noted that serum iron concentration and transferrin saturation may be higher in adolescent subjects than in mature women. Young adolescents have higher serum ferritin than older adolescents. Baily, Mahan, and Dimperio[5] illustrated the effect of less iron depletion during the early menstrual history as compared with mature women. Folicin deficiency may be more prevalent in the low-income pregnant population than iron deficiency.[5]

Therapeutic iron supplements and iron-rich foods are recommended, but complete iron studies to determine folacin levels and sickle preparations (for blacks) are indicated for patients with low hemoglobin or hematocrit levels.

3.4. Hypertension

One of the most important complications for adolescents is preeclampsia. In adolescent girls hypertension is rare, and documented hypertension is less than 1%. Preeclampsia is a hypertensive complication peculiar to pregnancy. It may be mild or severe, with or without convulsions or coma. There is no satisfactory etiology, and it occurs only during pregnancy. It is serious and common in adolescent pregnancy.[7] The first symptoms may be excessive weight gain, edema, and albumin in the urine before an elevation of blood pressure. Early discovery and institution of treatment includes reduction in the salt intake in the diet, and bed rest in the early stages is helpful for the teenager. Hospitalization is indicated in the more advanced cases. In teenage primigravida, increases of 30 mm systolic and 15 mm diastolic pressures are alarming and may be associated with convulsions and coma. Since the pathogenesis is unknown, treatment is symptomatic based on physiological principals and the use of drugs that will lower the blood pressure with the least likelihood of affecting the fetus. In primiparas, bed rest in the left-side position may promote diuresis and fall in blood pressure. A short labor and early delivery are desirable. A caesarean section can be utilized where indicated to shorten an otherwise prolonged labor. The presence of a neonatologist and an internist is manditory.

3.5. Preeclampsia

Preeclampsia appears to be a distortion of the normal physiological changes of pregnancy with vasoconstriction, antidiuresis, or increased interstitial fluid volume. Hypertension is associated with fibrin deposits in the glomeruli and hemorrhagic changes in the liver, brain, and other organs. Its clinical appearance becomes evident after the 24th week of pregnancy. Clinical and pathological studies indicate that the syndrome evolves gradually until the third trimester. Early diagnosis and management with bed rest and mild sedation may prevent severe progression of the disease. Antihypertensive drugs are indicated for elevated blood pressure, and delivery should be accomplished as soon as possible when eclampsia supervenes.

Ideal management requires a team approach by the obstetrician, internist, and neonatologist with a long-term follow-up. The preeclampsic teenager should be followed postpartum at 1-week intervals to maintain blood pressure and to continue medical evaluation and supervision.

3.6. Endocrine Disorders

A significant number of adolescents have simple endocrine disorders of the hypothalmus, pituitary, thyroid, parathyroid, adrenals, or ovaries. She should be referred to an endocrinologist.

3.7. Drugs

Pregnant teenagers take a great many drugs during pregnancy. They should be informed of the possible hazards of drugs during pregnancy including cigarettes and alcohol. Appropriate health history, continued observation, and prenatal teaching should help to reduce potential complications.

3.8. Abortion

Most available data support the position that abortion in the adolescent female has fewer long-range social and emotional consequences than does adolescent motherhood particularly if the teen keeps her child. The incidence of spontaneous abortion is unknown in the adolescent. Adolescents who abort with an IUD *in situ* should be considered septic. They should promptly be evaluated and antibiotics used as indicated.

3.9. Rubella and Cytomegaloviruses

A positive antibody titer before pregnancy affords protection against rubella. For women of childbearing age, immunization against rubella should be done only after susceptibility is established by a serologic test and negative pregnancy test.

Cytomegalovirus is often venereal and tends to be inapparent. The fetus may be infected during viremia of the mother or possibly by ascending virus-laden cervical secretions.

A study at Cleveland Metropolitan Hospital reported an infection rate of 1.2% or 1 in 83 infants. Infection was demonstrated in 60% of the mothers, all asymptomatic. Two-thirds of the women were primiparas and between 15 and 19 years of age. Infants may become infected in the birth canal. Infected IgM levels may be found in the cord blood or urine of the infant. Mothers and infants should be isolated from other patients.[6]

When fetal infection is clinically apparent, the congenital anomalies resemble those of rubella with infection during the second month or organogenesis resulting in more severe congenital anomalies. There is presently no way to prevent or treat cytomegalovirus infection.

Table I lists the effects of viral infection on the fetus and mother.

TABLE I. Viral Infections in Pregnancy[a]

Viruses	Maternal and/or fetal effects
Agents causing common respiratory illnesses	None
Influenza	Usually none; increased maternal morbidity and mortality in certain epidemics; fetal anomalies?
Mumps	Abortion; fetal anomalies?
Poliomyelitis	Increased maternal morbidity and mortality; congenital and neonatal paralytic poliomyelitis
Coxsackievirus B	Neonatal myocarditis and encephalitis; congenital heart disease?
Echovirus 6 and 9	None
Hepatitis	Abortion; stillbirth; prematurity; neonatal hepatitis; persistant postnatal infection (hepatitis B)
Measles	Abortion
Smallpox	Abortion; congenital disease
Vaccinia	Congenital disease
Varicella-zoster	Congenital disease; disseminated neonatal infection; congenital anomalies?
Herpes simplex	Prematurity; neonatal disease, localized or disseminated
Cytomegaloviruses	Prematurity; intrauterine growth retardation; congenital disease; congenital anomalies; persistant postnatal infection
Rubella	Abortion; stillbirth; intrauterine growth retardation; congenital disease; congenital anomalies; persistant postnatal infection

[a] From Horstmann.[7]

3.10. Diabetes

The Type I diabetic makes no insulin. Ideally the adolescent diabetic should be counseled regarding sexual relations and advised of her particular needs prior to the onset of sexual activity. The pill is not recommended, and the coil is sometimes not recommended due to the incidence of pelvic infection.

The pregnant diabetic should be advised to be in good control prior to the onset of a pregnancy. Good nutritional patterns that last a lifetime are recommended to reduce large swings in the glucose levels. Genetic counseling for the possibility of diabetes in the infant is recommended.

Fetal mortality varies from 20% and is generally three to six times greater than in nondiabetics. An unexplained increase in amniotic fluid occurs in diabetics. The infant of the diabetic mother is at risk for macrosomia hypoglycemia (10–25%), congenital anomalies (5–10%), respiratory distress syndrome (25–30%), hyperbilirubenemia (5–40%), and hypocalcemia (25%).[6]

The patient may be admitted to the hospital for an initial evaluation. Medical evaluation should include creatine clearance to check the level of kidney function, urine culture because diabetics are more prone to infection, and blood pressure because diabetics are more prone to hypertension and preeclampsia, and the blood sugar levels should be carefully monitored. The pregnant diabetic is prone to nocturnal and early morning hypoglycemia. Small changes in the insulin level may cause insulin shock.

An opthalmologist should be consulted for a baseline eye examination as diabetic retinopathy can increase during pregnancy.

Family support and teaching are essential to a good outcome. The nutritionist should work closely with the adolescent and family to maintain good eating habits. Closely supervised urine testing and home glucose monitoring, which helps maintain control between diet and exercise, can help maintain a patient at home. Blood sugars are done up to four times per day. For selected patients, an insulin pump, which gives small amounts of insulin for 24 hr, is advised.

The Type II diabetic makes insulin but may not make enough to maintain a person under the stress of pregnancy. An adolescent with a family history of diabetes should have a glucose tolerance test. Diet therapy is instituted if glucose levels are abnormal; insulin therapy, as described above, is necessary if the diabetes is not controlled by diet. The pregnant diabetic should be followed at least weekly.

3.11. Intrauterine Fetal Growth Retardation

There are several known and unknown causes of intrauterine fetal growth retardation resulting in low-birth weight infants who are small for gestational age. Adolescents, because of their low maternal age, some-times low socioeconomic status, possibly poor nutritional habits, smok-ing, and drug use, are at risk of having smaller-than-normal infants for gestational age. The cause may be fetal as well as chromosomal abnor-malities or congenital infections. Small placental size tends to produce small babies.[8] Severe prolonged preeclampsia, essential hypertension, renal hypertension, and vascular disease of advanced diabetes are con-ditions that exist when intrauterine fetal growth retardation is present.

Ultrasonic cephalometry can improve detection of intrauterine fetal growth retardation when performed repeatedly.[9] Improved nutrition with weekly nutritional evaluation, bed rest, and restriction of smoking and drug use are indicated.

4. PREVENTION

Adolescent pregnancy is in part the failure of society, the home, school, church, and health community to adequately teach sex education. Prevention begins with understanding and knowledge. People are partic-ularly sensitive and defensive about sex education. The physical body, attitudes, and powerful feelings aroused particularly in adolescence need to be discussed in the home and in the schools from the earliest time of a child's education. Sexual feelings are neither bad nor good. They are a part of the very essence of our being as are our sciences, history, phi-losophy, and fine arts.

Open communication between people that allows facts to be pre-sented, attitudes, fears, and anxieties to be expressed without put-down, and respectful answers to be given open pathways for understanding. Adolescents need other adults with whom to communicate. They know their parent's attitudes and discomforts and are exploring other view-points and putting to rest their own misunderstandings. Because this com-munication does not exist or exists with restraint, most teens learn about sexuality through their siblings or peers. When an adolescent chooses to become sexually active, it is a decision made on their own and not shared with a parent. Teens do not want to "hurt" their parents by having them feel that they have done something which their parents do not approve.

Much is appearing in the literature about the teaching of abstinence and values. There may be a question as to whether the place for such teaching is in a medical facility, where a patient comes for a contraceptive method. While parents may find it difficult to discuss contraceptive meth-ods with a teenager, as many in fact do, they are not opposed to or may even be grateful to have a school system teach family planning. Parents no doubt might accept additional education or counseling on how to teach their value system of sex to their own child. Early in the child's life, it is imperative to impart and give continued support to the philosophy of abstinence by theory and example. However, instruction should include cessation of abstaining and promote earlier and more effective contraceptive use by sexually active young peoole. Presently, eight in ten Americans favor sex education in the schools and seven in ten approve of birth control instruction. Many parents feel that they should and would like to teach their children about sexuality but feel unprepared to do so. Teens and their parents believe that sex education should be taught in the schools in cooperation with parents. The school certainly could provide accurate information and the home, the moral structure.

Only ten states and Washington, D.C., require or encourage sex education in junior and senior high school. Nineteen states have no policy requiring sex education. Many schools have fine curricula presently developed from kindergarten through high school but fail to implement it because of their own feelings concerning sexuality.

Physicians as part of the community can educate young people, parents, and educators in groups or individually in the office. Discussions should include anatomy, physiology, and methods of contraception (with emphasis on the consistent use of contraception if sexually active). The need for physical evaluation and sensitivity to the emotional concerns and the potential of disease processes and venereal diseases should also be discussed. Only one in seven adolescents initiates a visit to a family planning clinic before she becomes sexually active. Many (four in ten) come to a family planning clinic because they fear they might be pregnant.[2]

5. A HOSPITAL-BASED PROGRAM DESCRIPTION

5.1. Antepartum

The obstetrical teen clinic and family planning and outreach services at a large urban hospital and the teen mother clinic at a children's hospital provide comprehensive care and supplemental services to adolescents at risk of unintended pregnancies, pregnant adolescents, and adolescent parents and continued academic and vocation education, which are referral linkages predominantly with the academic and vocational public schools. Compared with adolescents who have not attended the obstetrical teen clinic, adolescents who have attended the teen clinic have a greater return rate to school and have an increased interval between the first and second pregnancy. Each girl receives close monitoring of her physical, educational, and social patterns in order to decrease maternal and fetal infant morbidity and mortality, and to improve nurturing patterns and her educational forecast. Observations also indicate that more teenagers who received individual counseling and support during pregnancy accepted contraception after delivery better than teenagers who had routine obstetrical care.

An appointment can be made at the regular obstetrical clinic by anyone who believes herself to be pregnant. The patient is interviewed for financial assessment and is referred for maternal infant care or referred to Medical Assistance for prenatal and hospital coverage if appropriate.

On the first visit, the patient is medically screened by the staff nurses, physician, and nurse midwife. At this time, a pregnancy test is done. Upon complete physical examination, including a pelvic examination, a gonococcal culture, Papanicolaou smear, and blood samples for VDRL (syphilis examination), complete blood count, blood type, Rh, and sickle preparation, are taken where applicable. If medical evaluation is needed by other specialists, an appointment is made for the patient. This is noted on the patient's chart, and a consultation sheet is returned when the patient has been evaluated. A patient may be referred for consultation at any time during pregnancy, and the above procedure is initiated. On the initial visit, the nurse explains the clinic procedures and basic comfort measures for pregnancy, answers questions, and provides the patient with reading materials, which are usually supplied by drug companies.

5.1.1. Option Counseling and Referral Services

Teen patients are offered all option available in the pregnancy on their initial contact with the obstetrical service by the social worker: parentlhood, adoption, foster care, or abortion. The focus of the teen clinic is on the preparation for parenthood, and patients initially planning adoption are followed in the regular obstetrical clinic, with a referral to appropriate adoption agencies. Adoption remains an option for teen clinic patients, and if that is chosen, the family is involved in the preparation. When an agency referral is made, the social worker acts as a liaison and support person in the hospital and follows the patient for resolution of the emotional aspects of relinquishing a baby. Few teen clinic patients choose adoption, but they are encouraged to consider their options.

Once the teenager has chosen to have and keep her baby, the social worker arranges an appointment to the teen clinic. Patients are seen every 2 weeks until 36 weeks gestation, then every week until delivery. Clinic visits are scheduled at closer intervals to allow for childbirth, socioeducation, and clinical supervision. At each clinic visit, the group is divided into two sections: a social group for the teens through their seventh month of pregnancy and a childbirth education class led by the nurse educator for teens in their eighth and ninth months of pregnancy.

5.1.2. Social Services

The function of the social worker and outreach worker is to help the teen patient use the resources available in the hospital and community to prepare herself, her partner, and her family for the care of the child. The prenatal period presents an opportune time to resolve family conflicts, learn information for developing decision-making skills, and form personal goals. The teen who is encouraged to learn responsible self-care during pregnancy has made a step toward competent parenthood. Referral for service not provided by the institution forms a comprehensive approach to the needs of the teen and her family.

5.1.3. Family Life Education

A series of 1-hour classes is incorporated into each teen clinic session to develop values and decision-making skills. Patients attend these classes through the seventh month of pregnancy. Classes are conducted by the social worker with assistance from the clinic and agency staff members. Classes include (1) an introduction to physical and emotional stages of pregnancy and how the pregnancy affects the self and significant others; (2) information on hospital and community resources, alternative education programs, and parent education programs; (3) birth control and veneral disease; (4) nutrition in pregnancy; (5) discussion of legal issues and personal values.

5.1.4. On-Going Clinic Social Services

The social worker and outreach worker are present at each clinic session to provide educational counseling and referral. Partners and family members are involved through the attendance at the clinic and scheduled counseling sessions. Services include: (1) the social worker and male outreach worker conduct a joint interview with each patient and her significant mate at the initial teen clinic visit and again in the eighth month of pregnancy to assess preparation for parenthood; (2) teens who have no supportive male partner are assisted in involving an appropriate family member; (3) the social worker acts as liaison for teen patients in foster care, residential placements, or those being followed by a social agency; (4) when necessary, referrals for intensive therapy are made to a child guidance clinic or community mental health center; (5) the social worker is available after classes for continued follow-up of individual problems.

5.1.5. Impact of Pregnancy on Educational Achievement

A teen clinic patient is expected to continue school during pregnancy or to make use of alternative educational and vocational resources. A frequent consequence of adolescent pregnancy is lowered educational achievement, with the life-long implications for personal and career satisfaction. Considerable energy by the teen clinic staff goes into helping the teen remain in the educational system.

In a hospital-based program, it is necessary to coordinate linkages to provide assistance with continuing education. Assessment of the school situation is a major part of the initial interview with the social worker. Services include: (1) The social worker acts as liaison between the clinic and school system for a teen attending school and assists drop-outs with reentry by individual and family counseling. (2) A referral to an alternative program for school-age parents is made at the initial interview if the patient desires or needs information on educational options. Many teens can slip through the loopholes of a system; therefore, the social worker facilitates utilization of available community resources. A referral and follow-up form are used to assure that teens do not drop out of school. Primary referral sources are schools that provide testing, a pregeneral education degree, a general education degree and vocational testing, and general education degree programs through the board of education. (3) Teens planning higher education are helped to discover options and identify resources through loans, grants, and special interest groups.

5.1.6. Provision for Day Care

When child care is not available through the family, day care is crucial to the success of educational or career plans. Day care is difficult at best to attain. In the prenatal period, exploration of day-care needs serves to prepare the teen for planning around a baby and for negotiating for help from her own or her boyfriend's family. Teens are encouraged to apply for day care early in their pregnancy due to long waiting lists at day-care facilities. Teens are also made aware of day-care resources, such as the associated day-care hotline. Continuing contact with the teen patient in the teen mother clinic includes assessment of changing day-care needs.

5.1.7. Childbirth Education

The nurse educator's goal is to clearly define to the adolescent the physical and emotional process she is experiencing during her pregnancy and prepare her for her labor and delivery experience, early parenting, and for future prevention of unwanted pregnancies. A series of eight classes are conducted for teens during their last 2 months of pregnancy. Partners or support partners are encouraged to attend. Antepartum care, fetal growth and development, labor and delivery (including a tour of the labor floor), anesthesia, caesarean section births, postpartum, family planning, and infant care and feeding are discussed. A pretest–posttest questionnaire is given so the teen staff and teens can assess their attitude and knowledge change at the end of their pregnancies (Fig. 1).

5.1.8. Individual Counseling

Each girl is counseled individually on hospital precedures (being on time for the clinic, what to do when appointments are missed, and where to come for emergency care). She is also counseled on ways to alleviate minor discomforts of pregnancy, sexual problems arising during pregnancy, sleep habits, weight gain, exercise, old wives tales, bowel problems, intake of vitamins and iron, travel, her right to a support person or partner during labor, and her right to "room-in" (have the baby in the room with her in the postpartum area) if she so chooses. Like all other members of the teen staff, the nurse educator makes clear to the adolescent that she is available for consultation any time the teen needs her.

5.1.9. Medical Care

Following the classes, the adolescents are seen medically. Because teens are at risk of preeclampsia, premature labor, nutritional anemia, intrauterine fetal growth retardation, and fetal and neonatal deaths, the teen obstetrical clinic sees patients every 2 weeks rather than at the usual 1-month intervals. Thus, the clinic hopes to have impact on the above-mentioned high-risk problems by providing close monitoring. At revisits, height, weight, and urinalysis are checked by the nurse; adequacy of uterine growth and related medical and nutritional problems are assessed by the physician and/or nurse midwife.

Each physician and the midwife follow their own patients. This gives continuity of care and helps develop a relationship of trust between the care provider and patient.

5.1.10. Services to Prospective Fathers

The male outreach worker provides individual counseling and job assistance to the partner of the pregnant teen, during the clinic period and in scheduled sessions.

Prospective fathers are invited to participate in the teen clinic through a letter, telephone, or personal contact with the social worker or outreach worker. At the first visit, the outreach worker conducts an interview to assess the family, work, and/or school situation. He provides counseling on these and relationship issues and is available at all times by telephone. Male partners are encouraged to participate in labor and delivery and to consider their role in the care of the baby. Postdelivery contact is made to support attendance in the family planning clinic, and the outreach worker is available following delivery.

Your response to these questions will help us to plan our education groups. Your answers are confidential. Thank you.

Dale Allison, Nurse Educator
Paige Slade, Social Worker

Name: _____

Date: _____

		True	False
1.	Menstruation is the shedding of the lining of the uterus which the body prepares for pregnancy each month.	____	____
2.	Pregnancy can occur at any time during the menstrual cycle.	____	____
3.	If you have your period, you know that you are not pregnant.	____	____
4.	The following words mean the same thing: birth control / family planning / pregnancy prevention / contraception	____	____
5.	If you are under 18, you need your parent's consent to get a birth control method from a doctor.	____	____
6.	Birth control pills are more dangerous to your health than pregnancy and delivery.	____	____
7.	The IUD or coil is a good method of birth control because it is always in place.	____	____
8.	Foam and condoms are good birth control methods which can be bought in any drugstore.	____	____
9.	The condom or rubber gives protection against V.D. because it prevents contact between the penis and vagina.	____	____
10.	You should only use birth control if your boyfriend wants you to.	____	____
11.	Boys are more experienced and will usually know how to keep a girl from getting pregnant.	____	____
12.	If you are under 18, you need your parent's consent for an abortion.	____	____
13.	Abortion cannot be done past 20 weeks of pregnancy.	____	____
14.	If you have an abortion, you may have trouble getting pregnant later.	____	____
15.	You can only get V.D. (venereal disease) from having sex with a person who has V.D.	____	____
16.	A woman can tell if she had V.D. because she will have signs such as pain or discharge.	____	____
17.	If you think you have V.D., you should douche a lot.	____	____
18.	When the signs of V.D. are gone, it means the man or woman is cured.	____	____
19.	If you are under 18, you must have your parents consent to be treated for V.D.	____	____
20.	Young women have more problems in pregnancy than women over 18.	____	____
21.	Bringing the baby's father to clinic will help him get ready to be a parent along with you.	____	____
22.	Men are not interested in pregnancy, but they will enjoy being fathers when the baby is born.	____	____
23.	Having sexual relations while pregnant will help feed the baby and make it grow.	____	____
24.	To relieve low back pain, wear low-heeled shoes, do pelvic rocking exercises, and have someone rub your back.	____	____
25.	A pregnant women should eat twice as much to feed the baby.	____	____

FIGURE 1. Teen clinic pregnancy questionnaire.

26. The baby can live outside the uterus after 6 months of pregnancy.   ____ ____
27. One can feel the baby move at 5 months of pregnancy.   ____ ____
28. If you have bleeding from the vagina, you should come to the emergency room when it stops.   ____ ____
29. Labor lasts 12–15 hr in a first pregnancy.   ____ ____
30. Labor is easier if you relax and take deep breaths.   ____ ____
31. My boyfriend should be with me in labor   1. to help me.
      2. to see the baby being born.   ____ ____
32. I can have my mother with me during labor and delivery.   ____ ____
33. If I call out loudly during labor and delivery,
   1. I will get help immediately.   ____ ____
   2. I will feel better.   ____ ____
   3. they will give me more pain medication.   ____ ____
   4. it will not help.   ____ ____
34. A caesarian section (C-section) is an operation to bring the baby out through the abdominal wall.   ____ ____
35. A C-section is done because:
   1. The head of the baby is too big for the mother's pelvis.   ____ ____
   2. The baby's heart rate drops.   ____ ____
   3. The afterbirth covers the opening of the cervix.   ____ ____
36. The placenta 1. is called the afterbirth.   ____ ____
      2. helps to nourish the baby in the uterus.   ____ ____
37. The small cut between the vagina and the rectum requiring stitches is called the episiotomy.   ____ ____
38. Good eating habits during pregnancy may prevent a premature birth.   ____ ____
39. What are the four food groups?   1. _____
      2. _____
      3. _____
      4. _____
40. Bottle feeding is better for a baby than breastfeeding.   ____ ____
41. During pregnancy, a woman should gain   1. less than 20 lbs.   ____ ____
      2. 22–30 lbs.   ____ ____
      3. 35–40 lbs.   ____ ____
Vitamins are   1. necessary parts of your diet.   ____ ____
   2. medicine.   ____ ____
   3. pills that may interfere with oral contraceptives.   ____ ____
43. A woman under 18 must have her parent's consent to place the baby for adoption or foster care.   ____ ____
44. Once the baby is born, a woman under 18 will get her own DPA grant.   ____ ____
45. It is up to the baby's father to provide money for the baby or not.   ____ ____
46. The baby's father has a right to visit the baby no matter what.   ____ ____
47. A blood test will tell who is the father of the baby.   ____ ____
48. If you are pregnant or have a baby, you may not be allowed to go to school.   ____ ____
49. Babies are hungry and eat a lot the first few days after birth.   ____ ____
50. Babies only sleep a little the first few days after birth.   ____ ____
51. Babies can be spoiled by holding and talking to them too much.   ____ ____
52. The best place for a baby to sleep is in the mother's bed.   ____ ____
53. I know all the answers to these questions.   ____ ____
54. I know a lot, but would like to learn more about my pregnancy.   ____ ____
55. Getting ready for parenthood is hard work!   ____ ____

FIGURE 1 (*Continued*).

5.1.11. Job Assistance for Male Teens

The outreach worker assists male partners of teen obstetrical clinic patients to discover and make use of job opportunities. Helping a young man to find a job not only enables him to provide financially for the baby, it also increases his self-esteem and the likelihood that he will move responsibly toward parenthood. The worker identifies areas in the hospital, university, and surrounding community where jobs may be available. He trains eligible young men in effective interviewing techniques and accompanies them on job interviews. Once a young man is employed, job progress is monitored through regular contact with the employer and employee. Through increased contact with the business community, the potential for jobs for teen partners grows.

The male outreach worker works with community and school personnel to involve partners of pregnant and never-pregnant teens. He expands his connections to the business community for the possibility of job openings and training programs. The outreach worker serves teen mothers in need of employment and coordinates service with the teen mother/pediatric clinic to assist partners in need of employment.

5.1.12. Nutrition Counseling

The nutritionist provides nutrition counseling for optimal nutrition of the adolescent and to decrease the high level of nutritional anemia seen in this population and counsels patients individually while they are waiting for the doctor. The nutritionist also has bimonthly group classes, incorporated into the social service groups. She discusses general nutrition for the pregnant mother and fetal development, iron needs during pregnancy, and intake needs during lactation.

The patient's family structure of shopping for food, preparation of food, and availability of refrigeration and cooking materials are determined through a personal interview. An assessment is made of the amount of food available to the patient, including if she is receiving a school lunch, the amount of energy expended daily, and the teen's capability for understanding verbal and written instructions. Supplemental liquid meals are provided by drug companies and prescription when necessary. Home visits, which assess the above-mentioned concerns, allow the health team to work with the teen's parents in the setting in which food is prepared.

5.2. Intrapartum and Early Postpartum

Teens are presently followed by the house staff and nursing staff during labor and delivery and the postpartum hospitilization. The nurse coordinator makes rounds on the postpartum floor to assess the teens postpartum progress, refers the patient to the family planning clinic, and provides continuity in the transfer of the patient to the teen family planning clinic and well-baby care. Classes in baby care and self-care are given during the postpartum hospitalization by the staff nurses.

The social worker assesses the teen in her preparation for parenthood and attachment to the baby. Her partner and family members are included in problem-solving counseling. The social worker explores plans for future contraception, education, and day care to provide necessary information and referral. After discharge from the hospital, the social worker is available to the teen patient by telephone and through the teen family planning clinic.

At 2 weeks postpartum, the patient is followed in the family planning clinic by the social worker and nurse midwife to assure consistency in use of family planning methods, return to school, and care of the newborn.

5.3. Family Planning Clinic

Special sessions of the family planning clinic are for adolescents. The average age of the teen presenting at the clinic is 14 years of age. A complete medical history is taken and all methods of contraception, including abstinence, are reviewed. The adolescent then attends a "rap session," which includes how to discuss sex and sexual feelings with partners and parents, family planning methods, sexual activity, venereal disease, and the pelvic examination.

Laboratory screening, weight, blood pressure, pregnancy testing, serology, and hematocrits are reviewed once a year.

Medical examination for the 2-week postpartum teen includes inspection of the episiotomy, discussion of general physical feeling, inspection of the caesarean section incision if necessary, instructions and permission on returning to school, and for well-baby appointments. (Some programs are able to combine well-baby visits with family planning visits.) A postpartum exercise program is recommended with a discussion on weight loss, pregnancy weight, and general hygiene. Referral of nongynecological medical problems is initiated, and information on breast-feeding is given as indicated. Contraceptive methods are provided. The postpartum teen is seen again in 4 weeks for a pelvic examination, which includes Papanicolaou and gonococcal smears.

Never-pregnant adolescents are also seen in the teen clinic and receive counseling as do the postpartum teens. A complete physical with a pelvic examination, including Papanicolaou and gonococcal smears, is given. Instruction in self-breast examination (and referral for other medical problems) is given, and hematocrit, serology, and contraceptive methods are prescribed.

Following the medical examination, the teen has an exit interview with the family planning counselor. At the exit interview, the counselor reviews the prescribed contraceptive method, makes necessary referral appointments, answers questions, gives instructions on therapeutic drugs for vaginal infections, and makes the next appointment.

The teen social worker provides continued follow-up of obstetrical teens for psychosocial concerns, school placement, and infant day care. The social worker is available to never-pregnant teens on a referral basis. Her partner, family members, or other helping persons are involved as needed to resolve problem situations. An outreach program to never-pregnant teens through the schools is previously described.

The health educator in the clinic maintains a 24-hr hotline for teens who are in need of immediate counseling. The family planning clinic in a medical facility is there to provide counseling for all medical services to the teens at risk of becoming pregnant.

5.4. Teen Tot Clinic

For the new teen mother and baby, new experiences in child development and care (parenting) are just beginning. Since 96% of teen mothers keep their children, and most of these children are cared for within the home, it is important for a medical center to provide physical care, immunization, and a parent education component. As a extension and in coordination with the program previously mentioned, the tot clinic was developed.

Teen mothers and fathers have the need and desire to learn effective parenting skills. The clinic is a natural place to provide parenting classes in conjunction with medical expertise.

5.5. Outreach Services

Adolescent programs containing health and educational components assist young parents to be better parents. However, for the sake of the teen's physical and emotional development plus career and economic stability, it would be more beneficial if childbearing were delayed past adolescence. Noting that there was an increased number of adolescents in the prenatal clinic, the obvious question was how the medical community with the support of the home, school, community, and media prevent unwanted pregnancies from occurring.

An eight-session program was developed and implemented in the junior high schools and high schools that served the same population as the hospital. The program includes a discussion of multiple-value systems, the importance of the ability to communicate about sex and sexuality, and how peer pressure affects values. Two sessions are devoted to anatomy and physiology and venereal diseases.

Pregnancy alternatives and birth control are presented through lecture and group discussion. The health educator from the university introduces the topic by emphasizing that there is no expectation that students are or should be sexually active. The facts on adolescent sexuality, pregnancy and abortion rates, common reasons why sexually active teens do not use contraception, laws pertaining to minors, and effective methods of family planning are presented. Symptons of pregnancy, pregnancy testing, alternatives to teenage pregnancy, single parenthood, implications of "forced marriage," adoption, and abortion continue the discussion. Emphasis is placed on the importance of seeking counseling when making such a decision.

The fifth session is a review of the topics previously covered with anomalous questions and evaluation by the students. The sixth class discusses pregnancy and its implications. The seventh class reviews personal responsibilities: unwanted pregnancies, the spread of venereal disease, sex and emotions, the need for considering feelings, and responsible decision making. The final class is a review and evaluation. It is made known that free family planning services are available to those who desire them.

6. Other Programs for Pregnant Adolescents

Programs regarding teenage pregnancy can be hospital based (previously discussed), school based, or community based. The physician in private practice should emphasize that a teen should remain in school throughout her pregnancy, return to school as soon as possible, and attend childbirth and parenting classes if possible and be referred for counseling if indicated.

6.1. School-Based Programs

Many students chose to remain in their own schools throughout pregnancy with medical passes for prenatal care. Others may feel more comfortable in attending an alternative school.

The school-based program is a setting where the teen can maintain learning at her own speed in her own subjects while she is pregnant. These schools often have childbirth and parenting classes plus social services to help with individual problems throughout pregnancy and in returning to the schools. A few schools have day-care centers for the infants of their student teens.

The school-based program is designed to provide direct education and social services and to insure that clients benefit from a truly comprehensive program in scope and time. In order to insure the provision of all services needed by pregnant adolescents and the young parents, the program coordinates with medical facilities, community legal service child development projects, day-care programs, and health advocacy programs. Social services are funded by Title XX of the Social Security Act. An example of this program is comprehensive services to school-age parents in Philadelphia. The program activities are divided into three major areas:

1. Alternative education classes for pregnant students and for young mothers. Students are eligible to begin in the seventh month of pregnancy. Classes are small and offer individualized academic instruction plus specialized health and home economics education related to pregnancy, childbirth, child development, and family planning. Social workers offer casework and group work services to the students, the babies' fathers, and their families. Several classes are geared to teach clerical skills and general education diploma preparation to young mothers who do not return to complete their high school education in a regular school.
2. Group work programs in regular high school, vocational schools, and special education centers for pregnant students, fathers, and mothers. These sessions are led by social workers once a week. The content includes information on pertinent topics and discussion designed to lead toward greater maturity and feeling of self-worth within the participants.
3. Follow-up services are offered by paraprofessional social workers for 2 years after the baby's birth. They are given mainly in home visits. Advocacy and help in securing social services are stressed. Information for coordination with appropriate health and social agencies is maintained.

6.2. Community-Based Programs

Some communities have set up comprehensive services to adolescents in churches and community Ys. They provide counseling, childbirth, and parenting classes and refer teens to local hospitals and schools for continuing health and educational services.

6.3. Programs to Enhance Early Childhood Education

The need for child care for infants and toddlers is enormous and the resources few. In order for a teenage mother to finish school, she has to have child care. At present this is primarily the responsibility of the adolescent's mother. Lacking day care, the chances for her preparing for self-support are minimal, the chance of recurrent pregnancies high. The long-range solution to the teenage pregnancy problem requires all the resources collectively and individually. Programs that have a child-development class for adolescents as part of their high school education have been developed. Day care in all settings must be applied for as early in the pregnancy as possible as the waiting lists are long.

6.4. Preparation of School Professional in Human Sexuality

School educators are constantly with adolescents, who are developing as sexual individuals, and yet choose to overlook this aspect of their education in favor of other material. Many schools have a well-developed human growth and development curriculum but fail to implement it because teachers feel ill at ease with the material.

Programs have been developed to help teachers feel comfortable with their own sexuality and the sexuality of their students and to help them feel comfortable in presenting information to their students, either as a group or individually.

Sex education would most appropriately be taught by teachers if teachers themselves were comfortable and knowledgeable about the subject. In Philadelphia, "Human Sexuality in a Changing Society" is a subject for the master's equivalency program. Its objectives are: (1) to increase basic knowledge of participants by presenting factual information on human sexuality and related subjects with special emphasis on how this knowledge may be applied to the teenager; (2) to stimulate thinking, share problems, and suggest solutions so that participants will be provided with a base for beginning work on sex education for the teenager; (3) to give participants the opportunity to present special problems related to sexuality of those with whom they are working (such as homosexuality, masturbation, venereal disease, and pregnancy), to share and examine solutions, and to learn of the available community resources to help them; (4) to provide available resource material with an evelation as to their authenticity and value; (5) to involve the participants in a process of self-evaluation on their own attitudes and feelings about sexuality that should enable them to relate more comfortably and realistically to their colleagues and the students and their parents; (6) to prepare each participant to emerge from the course with enough enthusiasm, confidence, determination, and direction to become a catalyst in his or her own setting for promoting and planning sex education programs.

6.5. Cure

We have considered the ills of teenage pregnancy and programs to help teens through a crisis in their lives and the lives of those who surround them. We face the question, "Is there a cure?" The medical profession's job is not to cure the ills of the society, it is within our realm to teach and prevent medical problems and to deal with them after they have arisen.

Implementing good sex education programs teaches youth about the biological and sexual aspects of their bodies and responsible relationships, and abstinence should always be stressed. Many people go through a lifetime, including childbearing experiences, without understanding sexuality. Teaching does not lead to promiscuous behavior.

In teaching methods of contraception to teenagers, avoidance of sex at the time of ovulation appears to be a less than responsible method. First, adults do not handle this method too well, and second, the irregularity of teens' menstrual cycles does not bode well for success. With the spontaneity of this age group for action, thoughtful planning leaves much to be desired in contraception by any method and especially with using nonovulatory time for activity. The consistency of the method should be taught as the prime factor, i.e., consistency in taking the pill, consistency in checking the IUD string, consistency in carrying foam, and/or especially consistency in being able to say "no."

Because sexual activity is a personal matter and cannot be legislated effectively, because teens have sexual feelings that are expressed through sexual activity, and because adolescents have partial development of their analytical probabilistic thought processes, some adolescent girls will continue to become pregnant. For the pregnant teen, their parents, and the infant-to-be, supportive services in health, education, preventative, and child care services should be available to help through the crisis of pregnancy and adolescence.

REFERENCES

1. The Alan Guttmacher Institute: *11 Million Teenagers*. New York, Planned Parenthood Federation of America, 1976, pp 7–12
2. The Alan Guttmacher Institute: *Teenager Pregnancy: The Problem That Hasn't Gone Away*. New York, Planned Parenthood Federation of America, 1981
3. Freeman EW, Rickels K: Adolescent contraception use: Current status of practice and research. *Obstet Gynecol* 53:338–394, 1979
4. Hendricks LE: *Unmarried Adolescent Fathers: Problems They Face and Ways They Cope with Them*, Final Report, Mental Health Research and Development Center, Institute for Urban Affairs and Research, Howard University, Washington, DC, September 1979
5. Baily LB, Mahan CS, Dimperio D: Folacin and iron status in low-income pregnant adolescents and mature women. *Am J Nutr* 33(9):1997–2001, September 1980
6. Kreutner A, Hollingsworth K, DR: *Adolescent Obstetrics and Gynecology*, Chicago, Yearbook Med Publ, 1978, pp 144–145 191
7. Horstmann DM: Viral infections, in Burrow, GN, Ferris, TF (eds): *Medical Complications during Pregnancy*. Philadelphia, Saunders, 1975

8. Molteni RA, Stys SG, Battaglia FC, Relationships of fetal and placental weight in human beings: Fetal/placental weight ratios at various gestational ages and birthweight distribution, *J. Reprod. Med.* 21:327–334, 1978
9. Quilligan EJ, Kretchmer N: *Fetal and Maternal Medicine.* New York, Wiley, 1980

BIBLIOGRAPHY

Audiovisual Resources

Films

The Facts of Life

Am I Normal?, Franklin Lakes, New Jersey, New Day Films,1979. 23 min, color.
About Sex, New York, Texture Films, 1972. 23 min, color.
A Baby Is Born: Highland Park, Illinois, Perennial Education, Inc.,1972. 23 min, color.

Sex Education—Parents

A Family Talks About Sex, EC Brown Foundation, Producer, Highland Park, Illinois, Perennial Education, Inc., Distributor, 1977. 29 min, color.

Teen Sexuality and Values

Are You Ready for Sex?, Highland Park, Illinois, Perennial Education, Inc.,1976. 24 min, color.

Teenage Pregnancy

And Baby Makes Two, Wilmette, Illinois, Films, Inc., 1978. 27 min, color. (Available from the Teenage Parenting Project of the Free Library, Philadelphia.)

Filmstrips

Pregnancy and Birth

Birth (Parenthood Series), White Plains, New York, Guidance Associates, 1977. Includes two filmstrips, two cassettes, discussion guide, script. Filmstrip 1—The Process of Birth, 21 min. Filmstrip 2—The Experience of Birth, 20 min.

Teenage Sexuality and Values

Sexual Values: A Matter of Responsibility, Pleasantville, New York, Sunburst Communications, 1978. Includes two filmstrips, two cassettes, discussion guide, script. Filmstrip 1—What's Important to You? Filmstrip 2—Making Sexual Decisions, 15 min.

Teenage Pregnancy

Pregnancy: A Teenage Epidemic? Wilton, Connecticut, Current Affairs Films, 1978. Includes: one filmstrip, one cassette, discussion guide. 17 min.

Becoming a Parent: The Emotional Impact (No. 1), Wilton, Connecticut, Current Affairs Films, 1978. Filmstrip 1—School-Age Parents—Who Are They? 10 min. Filmstrip 2—Social Pressures and Initial Decisions, 7 min. Filmstrip 3—Living Arrangements and Family Relationships, 7 min. Filmstrip 4—New Responsibilities, New Life Styles, 6 min. Filmstrip 5—Pursuing Personal Goals, 8 min.

Books and Pamphlets

The Facts of Life

Caring Mothers Cooperative: *Vaginal Hygiene.* Box 211, Tuckerton, New Jersey, 1978. 4 pp, $0.25.

Channing L Bete Co: *As You Grow Up.* Greenfield, Massachussetts, 1974. 15 pp, $0.50.

Channing L Bete Co: *What Every Woman Should Know about Rape.* Greenfield, Massachusetts, 1975. 15 pp, $0.50.

Choice: *Changes: You and Your Body.* 1501 Cherry St, Philadelphia, Pennsylvania, 58 pp, $1.00.

Hayes, MV: *A Boy Today . . . A Man Tomorrow.* Optimist International, 4494 Lindel Blvd, St. Louis, 1976. 16 pp, $0.50.

Institute for Family Research and Education: *Ten Heavy Facts about Sex.* Ed-U Press, 123 Fourth St, NW, Charlottesville, Virginia, 1975. 16 pp, $0.40.

Irwin, Theodore: *The Rights of Teenage as Patients.* Pamphlet No. 408. Public Affairs Committee, Inc., 381 Park Ave South, New York, 1972. 28 pp, $0.50.

Kimberly-Clark Corp: *Very Personally Yours.* Neenah, Wisconsin, 1976. 17 pp, $0.10.

Origins, Inc: *A Part of Our Lives,* 169 Boston St, Salem, Massachusetts, 1977. 27 pp, $3.00 ("A Part of our Live's" and "What Now? Under 18 and Pregnant," set, $4.50).

Personal Products Co: Esta's Crecienda Diuiertete. Box 69, Milltown, New Jersey, 1977. 26 pp, $0.10.

Personal Products Co: *For Boys: A Book about Girls.* Box 69, Milltown, New Jersey, 1977.

Planned Parenthood Center of Syracuse, Inc: *Growing Up* "*Specially for Pre-Teens and Young Teens.*" 1120 E Genesee St, Syracuse New York, 1973, 14 pp, $0.50.

Planned Parenthood Center of Syracuse, Inc: *Sex Facts.* 1120 E Genesee St, Syracuse, New York, 1977. 42 pp, $0.50.

Planned Parenthood Federation of America, Inc: *Sex Al-pha-bet.* 810 Seventh Ave, New York, 1973. 48 pp, $0.50.

Planned Parenthood Federation of America, Inc: *What Teens Want to Know but Don't Know How to Ask,* 810 Seventh Ave, New York, 1979. 13 pp, $0.50.

Planned Parenthood of Central South Carolina: *MASH: Mostly about Sex and Other Hassles.* Suite 202, 2719 Middleburg Dr, Columbia, South Carolina, 1975. 4 pp, $0.25.

Planned Parenthood-World Population: *Stop.* 3100 W Eighth St, Los Angeles, 1976. 8 pp, $0.20.

Population Services International: *The Man's World.* Suite 1019, 110 E 59 St, New York, 14 pp, $0.15.

Rocky Mountain Planned Parenthood: *The Perils of Puberty.* 1852 Vine St, Denver, 1974. 16 pp, $0.60.

Tepper, SS: *The Problems with Puberty.* Rocky Mountain Planned Parenthood, 1976. 20 pp, $0.60.

Tepper, SS: *This is You.* Rocky Mountain Planned Parenthood, 1852 Vine St, Denver, 1973. 9 pp, $0.50.

Pregnancy and Childbirth

Ashdown-Sharp P: *A Guide to Pregnancy and Parenthood for Women on Their Own*. Random House, New York, 1977.

Bean CA: Methods of Childbirth: *A Complete Guide to Childbirth Classes and Maternity Care*, Doubleday, New York, 1974.

Caplan F (ed): *The First Twelve Months of Life: Your Baby's Growth Month by Month*. Grosset & Dunlop, New York, 1973.

Caring Mothers Cooperative: *First Gifts*. Box 211, Tuckerton, New Jersey, 1978. 66 pp, $1.50.

Carson R: *Nine Months to Get Ready*, Pamphlet No. 376, Public Affairs Committee, Inc, 381 Park Ave South, New York, 1955. 22 pp, $0.50.

Gerber Products Co: *Expectant Mother's Guide*. Fremont, Michigan, 1978. 16 pp, free.

Goldbeck N: *As You Eat So Your Baby Grows: A Guide to Nutrition in Pregnancy*. Ceres Press, Box 87, Dept D, Woodstock, New York, 1978. 16 pp, $1.50.

Hall RE: *Nine Months Reading: A Medical Guide for Pregnant Women*, Revised edition, Bantam, New York, 1972.

Ksochnick K: *Having a Baby*. New Reader's Press, Syracuse, New York, 1975.

Maternity Center Association, *Comfort During Pregnancy*. 48 E 92 St, New York, 6 pp, $0.25.

Maternity Center Association: *Preparation for Childbearing*. 48 E 92 St, New York, 1977. 48 pp, $1.00.

Maternity Center Association: *Relaxation and Breathing*. 48 E 92 St, New York, 6 pp, $1.00.

Mientras Su Bebe Esta en Camino, Publication No. (OCD) 72-53 (S/N 017-091-0079-4). US Dept of Health, Education, and Welfare, Washington, DC, 1971, 29 pp, $0.45.

The National Foundation–March of Dimes: *Be Good to Your Baby before It is Born*. PO Box 2000, White Plains, New York, 1979. 16 pp, free. (Available in Spanish.)

The National Foundation–March of Dimes: *Food for Thought*. 1275 Mamaroneck Ave, White Plains, New York, 1978. 32 pp, 20/$1.50.

The National Foundation–March of Dimes: *Inside Your Body . . . Inside Your Head*. 1275 Mamaroneck Ave, White Plains Road, New York, 1978. 32 pp. 20/$5.00.

Pennsylvania Department of Health: *Pregnant? Be Nutrition Wise*. PO Box 90, Harrisburg, Pennsylvania, 1977. 6 pp, free.

Pennsylvania Department of Health: *What You Eat Makes a Difference*. PO Box 90, Harrisburg, Pennsylvania, 1978. 4 pp, free.

Phillips MG: *Food for the Teenager During Pregnancy*. US Department of Health, Education, and Welfare, Washington, DC, Publication No. (HSA) 77-5106 (S/N 017-026-00062-3), 1976. 24 pp, $1.20.

El Planeamiento Familiar y la Salud, Publication No. (HSA) 76-16028 (S/N 017-031-0014-7). US Dept of Health, Education, and Welfare, Washington, DC, 1977. 10 pp, $0.90.

Prenatal Care. Publication No. (OCD) 75-17 4(S/N 017-091-00187-1), US Department of Health, Education, and Welfare, Washington, DC, 1973. 70 pp, $1.05.

Procter and Gamble: *Care for Two: Baby and You*. Cincinnati, Ohio, 1972. 7 pp, free.

Pursel M: *A Look at Birth*, Lerner, Minneapolis, Minnesota, 1977.

So You're Going to be a New Father?, Publication No. (OCD) 73-28 (S/N 017-091-00190-1). US Department of Health, Education, and Welfare, 1973. 27 pp, $0.55.

Spock B: *Baby and Child Care*, Revised edition. Pocket Books, New York, 1976.

White BL: *The First Three Years of Life*. Avon Books, New York, 1975.

Sex Education—Parents

An Adolescent in Your Home. Publication No. (OHD) 77-30041 (S/N 017-091-00202-9) US Department of Health, Education, and Welfare, Washington, DC, 1976. 27 pp, $1.00.

Gordon S: *The Sexual Adolescent: Communication with Teenagers about Sex*. Duxbury, North Scituate, Massachusetts, 1978.

Gordon S, Dickman IR: *Sex Education: The Parents' Role*. (Pamphlet No. 549). Public Affairs Committee, Inc, 381 Park Ave South, New York, 1977. 28 pp, $0.50.

Hofstein S: *Talking to Preteenagers about Sex*. Pamphlet No. 476. Public Affairs Committee, Inc, 381 Park Ave South, New York, 1972. 24 pp, $0.50.

Landers A: *High School Sex and How to Deal with It: A Guide for Teens and Their Parents*, PO Box 11995. Chicago, 1977. 17 pp, $0.50.

Personal Products: *How Shall I tell My Daughter?* Box 69, Milltown, New Jersey, 1978. 13 pp, $0.08.

Planned Parenthood of America, Inc: *How to Talk to Your Teenagers about Something That's Not Easy to Talk about: Facts about the Facts of Life*. 810 Seventh Ave, New York, 1979. 22 pp, $0.25.

Planned Parenthood Center of Syracuse: *Sex Education at Home: A Guide for Parents*. 1120 E Genesee St, Syracuse, New York, 1974. 43 pp, $0.70.

Pomeroy WB: *Your Child and Sex: A Guide for Parents*. New York, Dell, 1974.

Sex Information and Education Council of the US: *Concerns of Parents about Sex Education*. (Study Guide No. 13). Human Science Press, 72 Fifth Ave, New York, 33 pp, $1.25.

Uslander AS, Weiss C, Telman J: *Sex Education for Today's Child: A Guide for Modern Parents*. Association Press, New York, 1977.

Teenage Sexuality and Values

Baker C: *Am I Parent Material?* National Alliance for Optional Parenthood, 2010 Massachusetts Ave, NW, Washington, DC, 1977. 5 pp, $0.10. (Available in Spanish.)

National Alliance for Optional Parenthood: *Are You Kidding Yourself?*, 1975. 8 pp, single copy free; two copies, $0.10.

CHAPTER 6

Contraceptive Use among Adolescents

George R. Huggins

1. INTRODUCTION

Successful contraceptive use among adolescents depends upon many factors. Medical, social, psychological, and economic issues all may exert independent effects on the young patient. Medical concerns are important in the choice of any therapeutic regimen, but they may be secondary or of minor importance in selection of a contraceptive method that an adolescent will utilize successfully. Successful contraceptive use depends upon resolution of both medical and nonmedical concerns of the patient.

With the adolescent, as opposed to the older patient, parental and family involvement must be considered. Parental counseling and understanding may play a most supportive and beneficial role in the decisions to begin or delay sexual activity. A well-informed parent may be the adolescent's best counselor when problems with a contraceptive method arise.

Unfortunately, many adolescents do not receive sexual counseling or education in the home. They are afraid and unable to discuss their questions, feelings, or fears with parents. They may not receive sex education in church or school either and therefore must obtain sexual and contraceptive information from informal and often misinformed sources, such as their peers.

1.1. Use of Contraceptives

Discussion of adolescent contraception presupposes sexual activity that may result in pregnancy. There is great concern about pregnancies

George R. Huggins • Department of Obstetrics and Gynecology, University of Pennsylvania Hospital, Philadelphia, Pennsylvania 19104.

that are unplanned or unwanted or occur out of wedlock. Of equal concern is the venereal disease that can result from this sexual activity. In one study,[1] the proportion of metropolitan women 15–19 years of age who had premarital intercourse rose from 30% in 1971 to 50% in 1979. By age 16, 20% of the women had had sexual intercourse—almost all premaritally. The percentage who had premarital pregnancies rose from 8.5 to 16.2% over the same time period, and 27% of those who became pregnant had never used any contraception. In another study[2] of approximately 1,142,000 pregnancies among teenagers in 1978, three-quarters were unintended. Of these, 434,000 were terminated by abortion. Other reports stated that only 192,000 children out of 1.1 million (one in six pregnancies) born to adolescent mothers had been conceived postmaritally.[3,4]

These same studies showed that even with a probable improvement in regular contraceptive use there has been no change in rates of premarital pregnancies for whites and only a slight decline for blacks. This is in contrast with the large decline in overall births rates in the United States since the 1960s. In fact, the 1979 survey showed some decreasing use of the most effective forms of contraception and an increase in use of less effective methods of use, such as withdrawal.

Adolescents who begin sexual intercourse are at high risk for unintended pregnancy, and this risk is higher for the younger adolescents, presumably because the younger women are less likely to use contraception during the early months after beginning sexual activity. Forty percent of women who began sexual activity before age 16 had not used contraception at first intercourse, while only 25% of 18- to 19-year-olds had not used contraceptives at first intercourse.[1]

Of women with premarital pregnancies who began sexual activity at age 15 or younger, 9% became pregnant within the first month of sexual activity and 20% had become pregnant within 6 months. For those women who waited until age 18–19 to begin sexual intercourse, these figures were approximately 5 and 11%, respectively.[5] Despite increased availability of nonprescription contraceptives and organized family planning programs, younger adolescents tend to delay seeking professional contraceptive help and use of effective contraception methods for more than a year after beginning sexual activity.[6,7]

Providing contraceptives to adolescents apparently has little or no effect on the incidence of premarital or extramarital sex relations. Adolescents rarely seek contraceptive assistance from their physician or family planning clinic until long after a pattern of sexual behavior has been established.[6]

1.2. Consequences of Early Adolescent Pregnancy

For adolescents whose pregnancies result in live births, the marital, educational, and socioeconomic consequences are devastating. In a 1971 study, 87% of mothers aged 15–19 kept their children in the household; 5% sent the infants to be cared for by family or friends; and 8% gave up the infants for adoption. Pregnancy is the single largest reason cited for school dropout among girls. Eighty percent of those who give birth at age 17 or younger never finish high school. These younger mothers are much more likely to come from poor families and are at high risk for unemployment and welfare dependency. A New York City study[8] showed that 91% of women who had first births at ages 15–17 were unemployed 19 months later. Seventy-two percent of these mothers were receiving welfare.

For married women whose first births occurred at age 17 or younger, 20% were divorced within 12 months and 60% were separated or divorced within 6 years of marriage. These rates were more than double those of women who delivered their first children at age 20 or older.[3]

When discussing adolescent pregnancies, one must be careful that all teens are not considered as a single group. Obviously there are marked differences between the 13-year-old nulliparous patient and a 19-year-old woman who has already had more than one child. The differences encompass psychological, socioeconomic, and psychosociological parameters. To counsel and advise the adolescent, her partner, or her family adequately, each person must be considered as a unique individual, and the multiple factors involved in successful contraceptive use must be applied to that unique person.

1.3. Reasons for Nonuse of Contraception

Among 15- to 19-year-old women, two-thirds of all pregnancies are unintended. However, among those women who do not want a pregnancy, 85% do not use contraception. Table I shows results of a study by Kantner

TABLE I. Percentage Distribution of Premaritally Pregnant Women Aged 15–19 in 1976 by Pregnancy Intention, Contraceptive Use, and Reasons for Nonuse[a]

Intention, use, and reason	Percent
Did not want pregnancy	100.0
Used contraception	14.5
Did not use	85.5
Reason for nonuse	
Didn't expect to have intercourse	43.0
Wanted to use something but couldn't in the circumstances	10.1
Partner objected	9.3
Believed it was wrong or dangerous to use contraception	12.5
Didn't know about contraception or where to get it	3.5
Sex wasn't much fun with contraception or contraception was too difficult to use	7.1

[a] Adapted from Ref. (9).

and Zelnik.[9] The most common reason given in this study for not using contraception was that the adolescent did not expect to have intercourse (43%). Less than 10% failed to use contraceptives because their partners objected.

The reasons teens give for delay in seeking professional help at family planning clinics are numerous and complex (see Fig. 1). The two most common reasons given were "didn't get around to it" (38%) and "afraid family would find out" (31%).

1.4. Effectiveness of Various Contraceptive Methods

Reports on the abilities of adolescents to continue to use any contraceptive method vary widely. Many patients who attend organized family planning programs tend to use oral contraceptives. Those who do not attend clinics tend to use condoms and withdrawal on a sporadic basis.

Early in counseling, it is best to determine what level of effectiveness is acceptable to the patient. Discussion about barrier methods or an intrauterine device (IUD) with a patient who feels strongly that the method of contraception must be virtually 100% effective is a waste of time. The patient must be counseled regarding the differences between theoretical effectiveness and use effectiveness. Oral contraceptives are more than 99% effective theoretically, but in any group of patients the use of effectiveness does not approach this theoretical figure. Indeed, in some populations the discontinuation rate for oral contraceptive use within the first year exceeds 50%. Unless these patients are provided with an acceptable alternative method of contraception, the pregnancy rates will be very high, not because of failure of the contraceptive method but because of failure of the patient to utilize the method adequately (see Table II). Adolescents who are motivated properly, well educated, and well counseled can use diaphragms with great effectiveness. Pregnancy rates as low as those with IUDs have been achieved.[11]

One commonly stated reason for discontinuation of birth control among adolescents is that they thought that their birth control method was dangerous. A major goal of counseling is to discuss the risks of each form of contraception, abortion, and term pregnancy with proper perspective for the patient.

Most contraceptive failures among adolescents occur during the early months of use. In our experience most discontinued use because of medical or subjective side effects from the method. With oral contraceptives the most common side effects were nausea, bleeding, breast tenderness, or breakthrough bleeding. With IUDs, side effects included pain and heavy or irregular bleeding. With barrier methods, users complained most commonly of messiness, loss of sensation, and irritation.[12,13] Increased frequency of follow-up visits and better supportive counseling can result in substantial improvement in continuation rates for all methods.

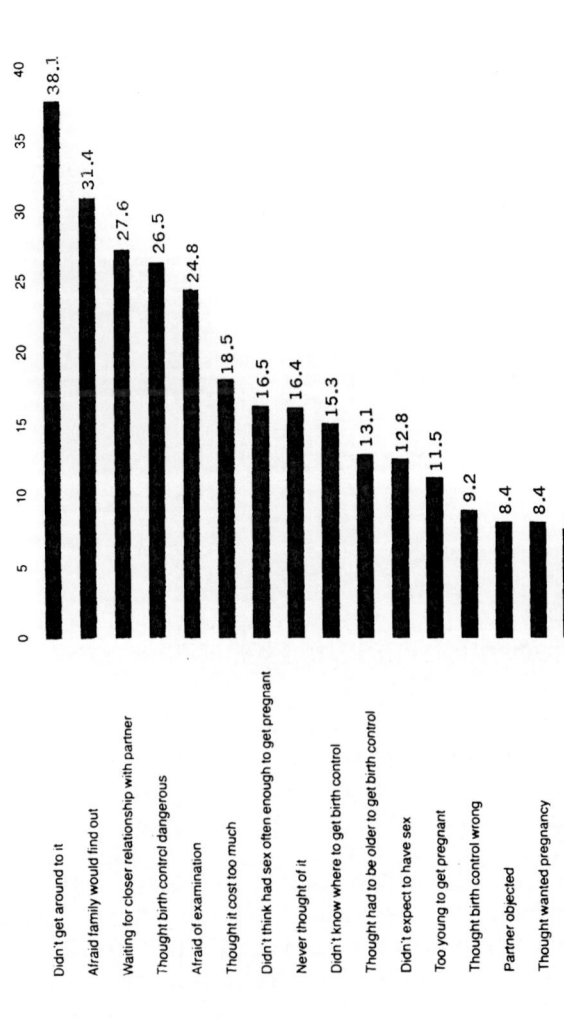

FIGURE 1. Reasons given by adolescents for delaying first visit to a family planning clinic. Note: Percentages do not add up to 100 because some patients gave more than one answer.

TABLE II. Method Effectiveness: Theoretical and Actual Use Rates Number of Pregnancies during the First Year of Use per 100 Nonsterile Women Initiating Method[a]

Method	Used correctly and consistently (theoretical effectiveness)	Average U.S. experience among 100 women who wanted no more children (actual use effectiveness)
Oral contraceptive (combined)	0.34	4–10
I.M. long-acting progestin	0.25	5–10
Condom and spermicidal agent	Less than 1	5–25 +
Low-dose oral progestin	1–1.5	5–10
IUD	1–3	5
Condom	3	10
Diaphragm (with spermicide)	3	17
Spermicidal foam	3	22
Spermicidal suppository	3	20–25
Coitus interruptus	9	20–25
Fertility awareness (natural family planning)		
Calendar only	13	21
Basal body temperature only	7	20
Cervical mucus only	2	25
Lactation for 12 months	15	40
Chance (sexually active for 12 months)	90	90
Douche	?	40

[a] Adapted from Ref. (10).

1.5. Adolescent Contraception and the Law

Young adolescents frequently ask family planning counselors, and less often private physicians, to provide contraception without parental notification or consent. This is a very complex and sensitive issue that may provoke considerable anxiety in the physician, related mainly to the old common law rule that consent of a parent or guardian is required before a doctor may provide any kind of medical treatment for a minor. An in-depth discussion of the legal aspects of this issue is beyond the scope of this chapter, but others have provided detailed reviews of this issue.[14,15] There has been a marked legal trend over the past few years to allow physicians and other contraceptive providers to render medical care to minors in areas regarding venereal disease, contraception, and pregnancy-related care without parental consent. Agencies that provide medical contraceptive services and that receive funds under Title XX (Social Services) and Title XIX (Medicaid) of the Social Security Act are required to provide such services to eligible persons, including "minors who can be considered to be sexually active," without regard to age or marital status.[14] Some states, e.g., New York and California, have statutes requiring that contraceptive services be provided to all needy persons, regardless of age or parental consent.[16]

Few issues in adolescent health care are as emotionally charged as this one. Parents may argue that the minor is their legal and moral responsibility and that they have a right to know about and be involved in contraceptive decisions. The adolescent, on the other hand, may argue for the right to privacy and the doctrine of patient–physician confidentiality. There is no easy resolution to these conflicting feelings. When faced with a request for confidentiality from a 16-year-old, the physician must weigh what is legal with what seems to be prudent and good medical practice. A physician is required legally to inform each user about the risks and benefits involved in the use of any contraceptive. In the case of the young adolescent, it is especially important to determine whether she is sufficiently intelligent and mature to understand the counseling and be able to weigh alternative choices. Informed consent is a critical component of adolescent health services, although it offers no easy solutions. As stated in the proceedings of the Conference on Ethical Issues in Adolescent Health:

> Informed consent from adolescents is problematic for all the same reasons adolescent sexuality is problematic. The in-between state of adolescence applies with respect to level of knowledge and ability to judge past experience, just as it applies to adolescent sexuality. Properly used, informed consent could serve as a pedagogical process. Adolescents could be helped to move from dependency of childhood, where other people make their decisions for them, to the independent and responsibile behavior of adulthood, where presumably they make some of their own decisions.[17]

It has been my practice for many years to counsel and encourage adolescents to involve their parents in decisions. Parental approval and support may have a marked beneficial effect on an adolescent's ability to use contraception successfully. Conversely, if the adolescent feels strongly that she cannot or does not wish to inform her parents and appears to understand the nature and consequences of her decisions to remain sexually active and to use contraception, her physician should provide her with contraceptive services in order to help prevent the almost inevitable unwanted pregnancy that results from unprotected intercourse.

On occasion, the physicians may be confronted with irate parents wanting to know why their daughter is being treated without their knowledge and consent. This most uncomfortable situation can be managed best in three separate counseling sessions, which can be achieved in one office visit.

The first session is with the parents, who should be encouraged to ventilate their frustrations, anger, and concern. These feelings can be focused on the physician, who is better equipped to handle them than the frightened daughter. Almost without exception, when confronted with the fact of a daughter's sexual activity, if the parents can discuss the alternative choices with a knowledgeable and supportive physician, they will agree to continued contraceptive use. The second session, with the adolescent, can be used to reassure her that she can now, without fear of reprisal, discuss her sexual activity and contraception with her parents. The third session should be a joint counseling session with the adolescent and parents. This may be the first opportunity they have had as a family to discuss openly the issues of sexuality, sexual behavior, and contraception. If managed properly these sessions may serve as the primary focus for a long overdue meaningful discussion between the adolescent and her parents.

In a completely opposite situation, the young adolescent who is just starting sexual activity or who is thought by her parents to be at risk of sexual involvement may find herself in a situation in which she is coerced by parents, teachers, social workers, or her physician into contraceptive use. The experience of a frightened 13-year-old who is brought to the doctor's office by a concerned mother demanding that the "child" be started on oral contraceptives is distressing to mother, daughter, and physician. In this kind of situation, it is imperative for the physician to determine from both the parent and the patient what circumstances elicited their concern. The mother may suspect erroneously that the daughter is sexually active. If this is so, the patient may be justifiably hostile to both mother and physician. In this atmosphere, no effective counseling or education can be accomplished. The physician must establish rapport and trust with both mother and daughter before there can be realistic hope of the patient's becoming a consistent and successful contraceptive user. It is critical to convince the adolescent that she, and not her mother, is your patient. She must feel secure that her discussions with her physician are privileged and respected. One soon realizes that a distrustful patient (whatever age) will not use effectively a contraceptive method that she feels has been forced upon her. On the other hand, the mother must be convinced that, although open unemotional communication among all concerned is most important and helpful, every patient needs to be secure in the knowledge that her wishes for confidentiality with her physician will be respected.

It is important that at the end of the first office visit the patient and her parents feel they have had adequate honest counseling and that they both have had input into two decisions:

1. The decision to contracept or not.
2. The decision that if the patient uses contraception, she feels comfortable that she is welcome and encouraged to call or return if any problems or questions arise.

1.6. Choice of Contraceptive Method

Contraceptive medical indications and contraindications for the adolescent are not substantially different from those in older patients. There are, however, several medical concerns that may assume more significance for the young patient. These are the effect of contraceptive method on future fertility, neoplasia, anemia, growth and development, and cardiovascular disease.

1.6.1. Future Fertility

The issue of fertility following contraceptive use is of the utmost importance to any adolescent who seeks advice about the "best" method of contraception. Few things are as disheartening to a couple and their physician as the discovery that the couple has an infertility problem after having postponed their childbearing functions for many years in order to achieve financial, social, or educational goals. The discovery of an unsuspected infertility problem often occurs in patients who have used long-term steroidal contraception, an IUD, or a combination of these.

Contraceptive usage today concentrates heavily on the utilization of oral contraceptives or IUDs. In addition to preventive contraceptive methods, more than 430,000 elective abortions were performed on teenagers in the United States in 1978.[18]

The return of spontaneous menses in patients who have discontinued oral contraceptives usually is prompt, occurring within 6–10 weeks in the vast majority of patients. Approximately 70% of patients ovulate during the first spontaneous cycle and 98% by the third cycle.[19] Because of a variable delay in the time of onset of the first cycle, patients must be cautioned not to expect these three cycles within the first 3 *lunar* months after discontinuing oral contraceptives. This return of menses does not appear to be age, dose, or time related.

Recently, researchers working on a large prospective study[20] reported delayed return to fertility in oral contraceptive users, compared with diaphragm users or IUD users. The delay was more marked in nulliparous women than in multiparous women. However, by 42 months after stopping contraception the pregnancy rates among the different groups were comparable, and there was no evidence of permanently impaired fertility. These nulliparous women were not adolescents, however, and there are no large prospective studies that address this issue specifically for teenagers.

As early as 1966, sporadic reports described the syndrome of prolonged secondary amenorrhea in patients who had taken oral contraceptives.[21] The cause of what has been referred to as "post-pill amenorrhea" has as yet to be established firmly. In ensuing years a number of observations have been made concerning this particular entity: (1) the "syndrome" does not appear to be related to any particular compound; (2) it is not time related, having been reported in patients who have taken oral contraceptives for as short a period of time as 3 months; and (3) it does not appear to be dose related, as it has been reported in patients taking various doses of both combined and sequential oral contraceptive preparations.[22] The incidence of secondary amenorrhea depends in part upon the particular time period the investigator uses before the patient is diagnosed as having post-pill amenorrhea. The incidence may be as high as 4% at 3 months, but drops to less than 1% at 1 year.[23] In some studies,[22] 35–50% of patients had experienced episodes of amenorrhea, menstrual irregularity, or late onset of menses prior to taking oral contraceptives. However, women with no previous menstrual abnormalities also have developed this particular syndrome.

If adolescents have experienced pregnancy or have well-established and regular menstrual patterns, there should be no concern that they will experience post-pill amenorrhea in excess of the expected incidence, regardless of chronologic age. For the very young nulligravidous patient whose menses are not regular there is some theoretical concern that taking oral contraceptives may mask an underlying gynecological endocrine problem or delay in some fashion the maturation of the functioning hypothalamic–pituitary–ovarian axis. However, such patients, if sexually active, are at high risk of pregnancy, and the slight risk of post-pill amenorrhea must be weighed against the high risk of pregnancy in the absence of effective contraception.

Discussion of treatment of post-pill amenorrhea is beyond the scope of this chapter. Data on therapy for teens are sparse, but the problem does not appear to be more difficult to treat in young patients. Indeed, because the patients are young, there is ample time to investigate thoroughly and use all available methods to resolve the problem.

1.6.2. Intrauterine Devices and Pelvic Inflammatory Disease

In a significant percentage of infertile patients there are tubo-ovarian, peritubal, or other associated pelvic adhesions that can be related to a recognizable episode or multiple episodes of prior pelvic inflammatory disease (PID). The incidence of PID and/or sexually transmitted disease (STD) is rare in women who are not sexually active. The incidences of gonococcal PID and nongonococcal STD have increased to epidemic proportions among sexually active teenagers. Age-specific rates for women under age 20 are much higher than are those for older women.[24] Two late sequelae of PID are infertility and ectopic pregnancy. Both of these have special importance to the teenager with no children or to those who have not completed their families. Ectopic pregnancies in the United States tripled in number from 1967 (13,200) to 1977 (41,000).[25] This increase parallels the increasing incidence of gonorrhea during the same time. The incidence of infertility from tubal occlusion increases with both the severity of the process and the number of infections. The range of infertility because of tubal occlusions postsalpingitis is shown in Table III.

Mishell et al.[26] showed that with the introduction of an IUD, each patient has her intrauterine cavity contaminated by bacteria. However, over a short period of time the uterus, through multiple biologic mechanisms, apparently is able to cleanse and sterilize its environment. The possible relationship between the use of an IUD and the development of PID has been studied extensively.[27,28] There is a widespread agreement in the epidemiologic literature that those who used IUDs had an increased risk of PID when compared with either noncontraceptors or with users of other contraceptive methods. The increased relative risk of PID varies from approximately 1.5 to 10.[27] Because most adolescents either are nulligravidous or have not completed childbearing, this increased risk is especially germane to them.

Attempts have been made to analyze risks of PID by both age and parity. Perhaps the most disturbing report with regard to adolescents was the study by Westrom et al. in 1976,[28] in which subjects had laparoscopy-confirmed acute salpingitis. The relative risk of acute salpingitis for women using IUDs was three times that in nonusers. For women using IUDs who had not been pregnant at any time, the corresponding risk was found to be sevenfold. This disturbing increased risk for nulligravidous women was not confirmed consistently in subsequent studies. The Women's Health Study, a large case-control study at 16 hospitals in nine cities across the United States, reported in 1981 that the relative risk of PID was significantly higher for nonblack women under the age of 25 compared with women over 25. However, there was not a significantly increased risk for nulliparous women.[29]

One problem inherent in these studies is the possible relative overreporting of diagnoses of PID. Approximately one-third of patients who have diagnosed PID as outpatients are found to have other diagnoses when investigated by laparoscopy. This has been the experience in our own institution by physicians at all stages of training and expertise. Without documenting the exact incidence of incorrect diagnoses, it should be ac-

TABLE III. Percentage of Infertility Postsalpingitis in a 15- to 24-Year Age Group[a]

Number of infections	Percent	Inflammatory changes	Percent
1	9.4	Mild	5.8
2	20.9	Moderately severe	10.8
3+	51.6	Severe	27.3

[a] Data adapted from Ref. (24).

cepted that our clinical acumen in diagnosing PID is not totally satisfactory.

Despite the consensus that there is an increased risk of PID among IUD users, the absolute degree of risk depends upon several factors. Age, sexual activity, marital status, number of sexual partners, and prior history of PID all have bearing on the incidence of this disease process. IUD use is only one of many factors that must be considered. Patients who do not have clinical episodes of PID and discontinue IUD use because of a desire for pregnancy achieve pregnancy rates within the same range as all women who attempt to achieve pregnancy. In the absence of clinical PID, short-term use (1–3 years) of an IUD does not appear to impair future fertility.[22]

Condom and diaphragm use appear to offer some barrier protective effect against PID. There also has been recent speculation that women who use oral contraceptives may have less risk of developing PID than do sexually active noncontraceptors.[30] For an adolescent with a history of one or more episodes of PID, these methods are the preferred contraception. Constant use of the condom will provide the highest degree of protection.

2. CONTRACEPTION AND NEOPLASIA

Because adolescents can be expected to use contraception for many years the question of potential carcinogenic effects from a contraceptive method is especially important.[31]

IUDs, barrier methods, and others: There is little suggestion that IUDs or other barrier methods are associated with development of either benign or malignant neoplasms. Patients may be reassured that with our present state of knowledge there is virtually no risk of inducting a neoplasm with these methods.

Oral contraceptives: Combined estrogen–progestogen oral contraceptives have been available in the United States since 1960. The first compound approved contained 10 mg of norethynodrel and 150 μg of mestranol. In the past 21 years, more than 30 brands of combined oral contraceptives have been sold in the United States. The steroidal dose has been reduced progressively. Most women are now using combined oral contraceptives with 0.3–1 mg of progestogen and 20–50 μg of estrogen. Since 1970, all of the newly introduced compounds have used ethinyl estradiol as the estrogen fraction, and all but one have used norethindrone as the progestogen. One compound contains 0.3 mg of norgestrel. Outside the United States, some of the 17 α-hydroxyprogesterone derivatives are in use. However, these derivatives were withdrawan from the United States market in response to concerns regarding the occurence of neoplastic lesions in dogs.

2.1. Problems of Investigation

The investigation of oral contraceptives and neoplasia is complicated by several problems: (1) there is no suitable animal model in most cases; (2) generally there is a long lag time, approximately 15 years following exposure to a carcinogen, until the development of overt malignancy; (3) there is a low incidence of malignant disease in the young female population; and (4) there are multiple etiologic influences, such as genetic, cultural, geographic, and environmental exposure to many possible carcinogens. Because of these problems the principal investigative methods in humans primarily involve various epidemiologic approaches. The methodologies used frequently are: (1) case reports (tumor registries), (2) disease rates and trends, (3) case-control studies, and (4) cohort studies. Each of these approaches provides a specific piece of a very complex puzzle. No one of these methods will provide a definitive answer about the causal relationships between exposure to an environmental carcinogen and the occurrence of disease.

2.2. Oral Contraceptives and Disease

Several major risk factors for carcinoma of the breast have been identified. The marked predisposition for this disease in women is attributed largely to the endocrine influence of estrogen on breast tissue. There is a genetic predisposition that is stronger in families with several relatives involved, especially if the disease has been premenopausal and bilateral. There is also a strong positive relationship with benign breast disease or precancerous mastopathy. This is most important because of the high incidence of benign breast disease in young women. In addition, the duration of preclinical or predetected breast cancer may extend over many years. If the growth of these tumors is stimulated by oral contraceptive steroids, one possible effect would be the appearance of lesions in younger women, and progression of the lesions, once clearly detected, would be enhanced. There are no firm epidemiologic data at present to support these concerns.

2.2.1. Breast Neoplasia

The relationship of combined oral contraceptives to benign breast disease seem to be consistently favorable. Most studies have shown oral contraceptive use to be associated with a significant *decreased* risk of benign breast neoplasia[22] Most studies have documented a decreased risk with increasing length of oral contraceptive use; this begins to appear after 2 years' use and persists for over 4 years. There is no suggestion that the risk increases with longer use or that there is an increased risk of benign or malignant disease after stopping oral contraceptives.

Despite the fact that oral contraceptives show a protective effect on benign disease, they do not show the same protective effect for breast cancer.[22] There may be at least two types of benign breast disease, one premalignant and the other not. The oral contraceptives appear to protect only against benign lesions that are not premalignant.[32] Current epidemiologic data show no association (either adverse or beneficial) between oral contraceptive use and the development of carcinoma of the breast in women.

2.2.2. Liver Tumors

Long-term combined oral contraceptive use appears to be related to the development of benign liver tumors. The risk of developing one of these rare lesions appears to be time, dose, and age related. The risk begins to increase appreciably after 4 years' continuous use, but women who are 27 years or older and who have used high-dose hormone pills for 7 years are at greatest risk for developing hepatocellular adenoma.[33] All of the patients described in the initial reports had been using mestranol-containing compounds. Subsequent analysis of case reports has continued to show a majority of patients who had taken mestranol-containing compounds, but significant numbers of these lesions have now been reported in patients who had taken compounds containing ethinyl estradiol. This may reflect prescribing bias, however, in that most of the early higher-dose pills contained mestranol, whereas the newer low-dose pills all contain ethinyl estradiol.

Although benign, these liver tumors may be life-threatening because of their tendency to spontaneously hemorrhage, with resultant death. Fortunately these lesions are quite rare—the estimated incidence is 1 to 5 per 1,000,000 women. There is some evidence that these lesions regress spontaneously following discontinuation of the oral contraceptive.[22]

The standard liver function tests are of little value in diagnosing these lesions, as no consistent abnormalities have been found. Celiac arteriography, ultrasonography, and liver scanning offer the most accurate diagnostic approaches. These lesions may be detected initially as asymptomatic masses, but the majority of patients complain initially of epigastric or right upper-quadrant abdominal pain.

All adolescents who have been taking oral contraceptives for 4 or more years should be cautioned to report symptoms of abdominal pain or distress. They should be examined abdominally for evidence of liver enlargement every 6 months. Consideration should be given to stopping the oral contraceptive after 4 years for a 1-year period if the patient can reasonably be expected to adequately use another effective method of contraception.

2.2.3. Endometrial Hyperplasia

Sequential oral contraceptives, which involved the administration of estrogen for 14–16 days followed by an estrogen–progestogen combination for 5–6 days, were marketed first in the United States in 1963. By 1976, there were 20 cases of verified invasive carcinoma of the endometrium in women under age 40 who had documented use of sequential type preparations. Following these reports the sequential oral contraceptives were removed from the market voluntarily by the manufacturers. There is no suggestion that combined oral contraceptives or progestogen-only pills contribute to neoplasia. In fact, they seem to exert protective effect against endometrial hyperplasia and may be used therapeutically to treat menorrhagia in teens.

2.2.4. Cervical Neoplasia

The epidemiologic factors responsible for the development of cervical dysplasia and subsequent progression to carcinoma in some patients are complex and poorly understood. Two of the major epidemiologic factors in carcinoma of the cervix seem to be young age at first intercourse and multiple sexual partners. There is also the confounding influence of the possible carcinogenic potential of the sexually transmitted herpes Type II virus.[22] Certainly the woman who begins sexual activity at a young age and who has multiple partners is at increased risk to contact genital herpes infection. These factors place sexually active unmarried adolescents at increased risk of cervical dysplasia and subsequent development of carcinoma of the cervix. In addition, these factors confound many of the studies that have been attempted to study the effect of oral contraceptives on the development of dysplasia or carcinoma. Despite large numbers of both case-control and cohort studies, only one epidemiologic study before 1975 specifically analyzed the issues of age at first intercourse and contraceptive choice.[34]

Most studies have shown no increased incidence of dysplasia or carcinoma *in situ* among oral contraceptive users, compared with users of other forms of contraception. However, concern persists regarding long-term use of oral contraceptives among women who start using them in their early teens.[35] There is some suggestion that long-term use is associated independently with increased risk of dysplasia and carcinoma *in situ*; the risk is highest among women who had used oral contraceptives for 10 years or longer.[36] These concerns require further study.

Condom and diaphragm use may exert some barrier protective effect against STDs. Comparisons between incidence rates of infection and cervical neoplasia in diaphragm and condom users versus oral contraceptive users shows a lower rate among diaphragm users. This may demonstrate a protective effect of diaphragm and condom use rather than an increased rate with oral contraceptive use.

2.2.5. Pituitary Tumors

The association of post-pill amenorrhea, galactorrhea, and the existence of pituitary adenomas has been described. The present data are somewhat conflicting, and no firm conclusions can be made regarding an association between pituitary adenoma and oral contraceptive use.[22]

2.2.6. Ovarian Neoplasms

Several studies have reported a lower incidence of functional ovarian cysts in oral contraceptive users than in nonusers. There have been few epidemiologic studies concerned with ovarian cancer and oral contraceptive use. The data presented are reassuring (although incomplete), demonstrating no increased incidence and no association between ovarian cancer and oral contraceptive use. One recent case-comparison study of 300 women with ovarian cancer suggested that oral contraceptive use might protect against ovarian cancer.[37] These findings require confirmation from other diversified patient populations.

2.3. Other Medical Concerns

2.3.1. Anemia

IUD use increases the occurrence of abnormal uterine bleeding. Menses may be heavier than expected and the total time of menstruation may be prolonged. In addition, a significant number of patients will experience bleeding at times other than their menses. This metrorrhagia is more common in the month following IUD insertion and has a tendency to diminish over the succeeding few months. Excessive, prolonged, or irregular bleeding is one of the most frequent symptoms leading to discontinuation of the IUD.[38] The adolescent who may already have low iron stores is particularly vulnerable to increased blood loss that may be IUD related. Both the copper-bearing and inert devices have been shown to be associated with increased menstrual blood loss. The increased loss averages 50–100%. Eventual iron depletion and anemia from this blood loss may occur in 30–50% of women who use IUDs.[39] This is not true, however, with the new progestogen-releasing IUDs. Menstrual blood loss has been shown to decrease after insertion of these devices.[40] Any adolescent using an IUD should have her hemoglobin and hematocrit levels checked periodically, as anemia may develop in patients who may not perceive increased blood loss.

Combined oral contraceptives have several effects on menstruation.

In the first month of use, they may be associated with an increased incidence of metrorrhagia, or breakthrough bleeding. This may occur in more than 10% of patients. Over the first 3 months of use, this incidence should be less than 5%. However, the average total blood loss for women taking oral contraceptives is decreased and, indeed, for patients taking low-dose estrogen pills, hypomenorrhea and amenorrhea may result. This may be a beneficial side effect for adolescent patients who are anemic or who have heavy menstrual periods.

Young nulligravidous women experience other IUD-related problems in higher proportions than those who are multiparous. In general, although the pregnancy rates with the CU_7, CUT, and Progestosert are comparable with multiparas, the expulsion rates and removals for medical reasons, mostly bleeding, are higher among those who have never been pregnant.

2.3.2. Abortion—Immediate and Delayed Risks

The 434,000 adolescents who underwent induced abortions in 1978 were subjected to immediate medical risks and possible delayed risks to their future fertility. In 1978, 1,157,776 legally induced abortions were reported to the Centers for Disease Control. Of these, 79% occurred at less than 10 weeks' gestation, 12% at 11–12 weeks, and 9% at greater than 12 weeks.[18]

The immediate risks of major abortion-related morbidity, such as hemorrhage, uterine perforation, cervical injury, infection, and death are related directly to the duration of pregnancy and type of procedure performed (see Fig. 2). The morbidity and mortality rates of an early first trimester abortion are quite low in relation to morbidity and mortality related to a full-term pregnancy. However, the rates rise steadily, so that the risks of late second trimester terminations equal or exceed the risks of a full-term delivery.

Pregnant adolescents are more likely than women over the age of 20 to delay decisions about obtaining an abortion. This delay results in later terminations with resulting increased immediate morbidity.

2.3.3. Future Fertility and Abortion

There are conflicting reports regarding the effect of therapeutic abortion on future fertility. Special attention has been given to (1) the incompetent cervical os and increased prematurity rates, (2) pelvic adhesions resulting from postoperative infection, (3) Rh sensitization, and (4) the development of Asherman syndrome (the presence of intrauterine synechiae that produce clinical symptoms, i.e., menstrual abnormalities, infertility, habitual abortion).

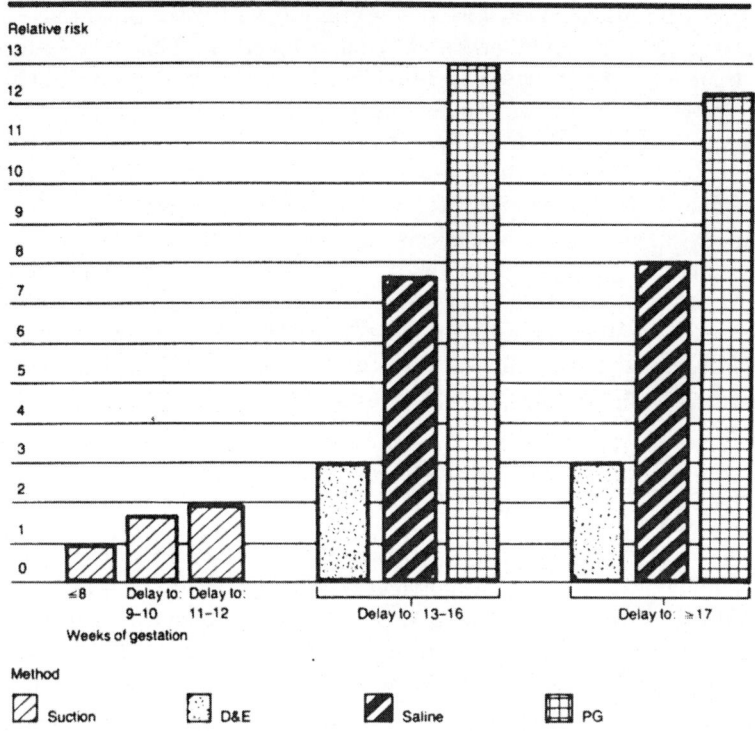

FIGURE 2. Relative risk of major abortion-related morbidity due to length of gestation and choice of method, compared with risk associated with suction ≤8 weeks' gestation.

2.3.4. Incompetent Cervix

The World Health Organization task force on the late sequelae of abortion[42] concluded tentatively that centers using small caliber dilitation and suction showed no increased incidence of spontaneous abortion, whereas those doing late abortions with larger caliber dilitation and sharp curettage did show some increase in subsequent pregnancy wastage. As with early complications, occurrence of late complications and effects on subsequent pregnancy appear to be related directly to the length of gestation.

2.3.5. Postoperative Infection

Postoperative infection, most often related to incomplete evacuation of tissue, may produce pelvic adhesions, increasing the incidence of infertility and ectopic pregnancy. This infection may be mild and self-limiting and may go unnoticed by the patient or her physician. Persistent temperature elevation greater than 38°C is one tentative indication of postoperative infection. The incidence of infection in patients who have had abortions by suction and curettage at less than 12 weeks' gestation is less than 1%, while the incidences for saline and prostaglandin-induced abortions are 5 and 11%, respectively.[42]

It is impossible with present data to document the extent of infertility and ectopic pregnancy subsequent to induced abortion.[41] However, it is reasonable to assume that their occurrence in patients who become clinically infected postabortion would be similar to that seen in other patients who develop PID.

2.3.6. Transplacental Hemorrhage and Rh Sensitization

As many as 5–10% of Rh-negative women who have induced abortions may become sensitized. This risk is proportional to the length of gestation. It is virtually nil at 1 month, 2% at 2 months, 9% at 3 months, and 12% for second trimester terminations.[42] This sensitization may be prevented effectively by the administration of Rh immunoglobulin postoperatively. For women with first trimester abortions, a 50-μg dose is sufficient. For women undergoing second trimester terminations, a 300-μg dose, the same as the postdelivery dose, should be administered.

2.3.7. Asherman Syndrome

In the past, Asherman syndrome has been relatively rare and, when described, has had a high correlation with preexisting intrauterine infection associated with vigorous dilitation and curettage. However, there have been recent reports of Asherman syndrome in patients who have undergone early elective first trimester termination of pregnancy without immediate postoperative complications. Management of this problem involves surgical intrauterine lysis of the adhesions, followed by postoperative administration of estrogens. Return of menses correlates poorly with establishment of a pregnancy that can continue successfully to the term birth of a live infant. Live birth rates of 20–50% may be expected in such cases.[22]

Unfortunately, increasing numbers of women are electing repeat abortion as a primary means of birth control. In 1974, 13% of women who obtained legal abortions in the United States had had at least one previous abortion. By 1976, this had increased to 23%.[43] Adolescents who have one abortion, because of their long subsequent reproductive lives, are more likely to be subjected to repeat abortion, and the risks in these patients are compounded with each succeeding operation.

2.3.8. Vascular Disease

Oral contraceptive use has been associated with various forms of cardiovascular disease. Hypertension, thrombophlebitis, pulmonary embolism, cerebral thrombosis, cerebral hemorrhage, and myocardial infarction all have been reported.

Hypertension has been reported in 1–19% of previously normotensive women who were taking oral contraceptives.[44] This has included every type of hypertension, from mild to severe. In general, the reported rise in blood pressure is mild. It may develop after short-term use (1–3 months) but is more common in women after taking oral contraceptives for 2–4 years. Stopping oral contraceptive use usually is followed by a slow return to normotensive levels.

There is no evidence that adolescents are more prone to develop oral contraceptive-induced hypertension than are women in their twenties, and adolescents may be less prone than women in their thirties. Neither is there firm evidence to show that adolescents who take oral contraceptives are more prone to develop hypertension at older ages. An adolescent who uses oral contraceptives should have blood pressure levels taken every 6 months. Any significant elevation in pressure should be investigated promptly.

The other serious sequelae of oral contraceptive use, pulmonary and cerebral embolism, cerebral hemorrhage, and myocardial infarction, are rare and occur more commonly in women older than 35. The adolescent may be considered to be at low risk for developing these serious problems and there is no firm evidence to show that use of oral contraceptives in adolescents increases the risk of developing these problems in later years.

2.3.9. Growth and Development

There has been some concern that oral contraceptive use among young adolescents might induce premature epiphyseal closure, thereby retarding normal growth. The ovaries produce substantial endogenous estrogen for some time prior to menarche, when the epiphyseal plates have begun to close. Administration of oral contraceptives in adolescents who have established a menstrual pattern will not accelerate this closure rate. There should be no concern that oral contraceptives will retard growth in these patients.

3. METHODS TO IMPROVE CONTRACEPTIVE UTILIZATION

Despite the complex psychosocial nature of adolescent behavior and a lack of consensus on one most effective program methodology, there are some approaches that may improve adolescent contraceptive use.[45]

3.1. Increased Sex Education

Because there is little or no knowledge about sexuality and reproduction among a substantial proportion of adolescents, there is a need to
institute sex education in the home as well as in the schools. This should
be done at early ages. Waiting until high school is too late for many teens.
Family planning programs serve little purpose at present in reaching adolescents before they begin sexual activity. The major role of these programs appears to be one of information, support, and encouragement to
those who are already sexually active. These efforts may be improved by
establishing liasons with school-based educational programs. In 1976, the
St. Paul Maternal and Infant Care Project established health clinics, including contraceptive care in two high schools. In the subsequent 3 years,
25% of the female students used the family planning service; 86% continued contraception for 3 years. The pregnancy rate declined by 40%.[46]

Organizing small group discussions or "rap sessions" helps to encourage discussion of many sensitive issues and helps the adolescents
organize better their analytic thought process. Patients also are helped to
connect information and behavior, and this may lead to better continuation of contraceptives.

3.2. Family Planning Programs

Increasing access to care will increase substantially the number of
sexually active adolescents who use these services. Approximately
214,000 women aged 15–19 attended organized family planning projects
in 1969. During the early 1970s, this number increased dramatically, with
the rapid expansion of family planning services in both the public and
private sectors of the United States. In 1975, 1.2 million women under
age 20 received contraceptive services from family planning clinics. These
women comprised 30% of the total patient population seen in organized
family planning programs.

3.3. Role of the Private Physician

Private physicians are responsible for less than 20% of the contraceptive services provided to adolescents in the United States. This is in
marked contrast with adult female health care in which the norm of the
heatlh care delivery system is the private physician from whom a majority
of women obtain contraceptive services.[45] Certainly these physicians
could play significantly increased roles at least with the sons and daugh-
ters of their own patients. Both private physicians and family planning

clinics can improve adolescent contraceptive care by the following:

1. Consider each patient's individual needs.
2. Insure adequate education.
3. Involve both patient and family in decisions.
4. Assume the patient's voluntary acceptance of the contraceptive method.
5. Provide firm guidance about the decision to avoid pregnancy and the method chosen.
6. Provide emotional and medical support for patients who have side effects.
7. Be willing to allow the patient to experiment and change methods under medical supervision.

4. RECOMMENDATIONS FOR CONTRACEPTIVE USE

4.1. Oral Contraceptives

1. May be prescribed without regard to age or parity once menstrual pattern has been established.
2. Preferred for patients who are anemic or have severe menorrhagia or severe dysmenorrhea.
3. See Table IV for contraindications.

4.2. IUDs

1. Recommended where appropriate if the patient has experienced a pregnancy that has concluded either with delivery or abortion, regardless of patient's age.

TABLE IV. Contraindications against Combined Oral Contraceptives

Absolute	Strong relative
History or presence of	
Thromboembolic disease	Vascular or migraine headaches
Cerebrovascular accident	Hypertension
Coronary artery disease	Diabetes
Hepatic adenoma	Gallbladder disease
Breast malignancy	Sickle cell disease
Presence of	
Pregnancy	Mononucleosis—acute phase
Impaired liver function	Long leg casts
Active liver disease	Severe renal disease
Undiagnosed abnormal uterine bleeding	

TABLE V. Contraindications against IUD Use

Absolute	Relative
Active pelvic infection	Recent or recurrent PID
Pregnancy	Acute cervicitis
Bicornuate uterus	History of ectopic pregnancy
Undiagnosed abnormal uterine bleeding	Valvular heart disease
	Clotting disorders
	Allergy to copper (copper-bearing device)
	Small uterus
	Severe menorrhagia
	Severe dysmenorrhea

2. If the patient has never been pregnant, has normal menses, does not have severe dysmenorrhea, and has a normal-size uterus, an IUD can be used. The physician should recognize that younger adolescents have higher rates of expulsion and medical removal because of side effects.
3. If the patient has never been pregnant or has menorrhagia, dysmenorrhea, or a small uterus, IUD use is discouraged strongly.
4. See Table V for containdications.

4.3. Barrier Methods

Condoms and diaphragms are recommended for:

1. Patients who have contraindications against use of oral contraceptives or IUDs.
2. As a backup method for patients just starting to use oral contraceptives or IUDs.
3. Patients who request them.
4. Patients who have infrequent intercourse.

Spermicides are recommended for the same reasons as barrier methods. Rhythm and abstinence are recommended only for those patients who are strongly motivated and who are comfortable handling their bodies and have regular menses.

REFERENCES

1. Zelnik M, Kantner JF: Sexual activity, contraceptive use and pregnancy among metropolitan teenagers: 1971–1979. *Fam Plan Perspect* 12:230–237, 1980

2. Henshaw S, Forrest JD, Sullivan E, et al: Abortion in the United States 1978-1979. *Fam Plan Perspect* 13:136, 1981

3. Anonymous: *Teenage Pregnancy: The Problem That Hasn't Gone Away.* New York, The Allan Guttmacher Institute, 1981

4. Annonymous: *11 Million Teenagers. What Can Be Done about the Epidemic of Adolescent Pregnancies in the United States?* New York, The Allan Guttmacher Institute, 1976

5. Zabin LS, Kantner SF, Zelnik M: The risk of adolescent pregnancy in the first months of intercourse. *Fam Plan Perspect* 11:215–222, 1979

6. Akpom GA, Akpom KL, Davis M: Prior sexual behavior of teenagers attending rap sessions for the first time. *Fam Plan Perspect* 8:203, 1976

7. Settlage DSF, Boroff S, Cooper D: Sexual experience of younger teenage girls seeking contraceptive assistance for the first time. *Fam Plan Perspect* 5:223, 1973

8. Presser HB: Social consequences of teenage childbearing, in Petersen, W, Day, L (eds): *Social Demography: The State of the Art.* Cambridge, Harvard Univ Press, 1977

9. Zelnik M, Kantner JF: Reasons for non use of contraception by sexually active women aged 15-19. *Fam Plan Perspect* 11:289–296, 1979

10. Hatcher RA, Stewart GK, Stewart F, et al: *Contraceptive Technology 1980–81,* ed 10. New York, Irving Publishers, 1981, p 4

11. Lane M, Arceo R, Sobrero JA: Successful use of the diaphragm and jelly by a young population: Report of a clinical study. *Fam Plan Perspect* 8:81–85, 1976

12. Mudd EH, Dickens HO, Garcia C-R, et al: Adolescent health services and contraceptive use. *Am J Orthopsychiatr* 48:495–504, 1978

13. Huggins GR: Counseling patients for contraception. *Clin Obstet Gynecol* 22:509–520, 1979

14. Paul EW, Pilpel HF, Wechster NF: Pregnancy, teenager and the law, 1976. *Fam Plan Perspect* 8:16–32, 1976

15. Paul EW, Pilpel HF: Teenagers and pregnancy: The law in 1979. *Fam Plan Perspect* 11:297–302, 1979

16. New York Services Law 131-e and Department Social Services Reg.s. 386. 16, Cal. Welfare and Inst. Code Ss 10053, 2

17. American Public Health Association: *Ethical Issues in Adolescent Health, Proceedings, December 6–7, 1976.* Washington, DC, American Public Health Association

18. Centers for Disease Control: *Abortion Surveillance 1978.* Atlanta, DHEW Publication, November 1980

19. Larsson-Cohn U: The length of the first three menstrual cycles after combined oral contraceptive treatment. *Acta Obstel Gynecol Scand* 48:416, 1969

20. Vessey MP, Wright NH, McPherson K, et al: Fertility after stopping different methods of contraception. *Br Med J* 1:265–267, 1978

21. Shearman RP: Amenorrhea after treatment with oral contraceptives. *Lancet* 2:1110–1111, 1966

22. Huggins GR: Contraceptive use and subsequent fertility. *Fertil Steril* 28:603–612, 1977

23. Petterson F, Fries H, Nillius JS: Epidemiology of secondary amenorrhea. *Am J Obstet Gynecol* 117:80, 1973

24. Westrom L: Incidence, prevalence and trends of acute pelvic inflammatory disease and its consequences in industrialized countries. *Am J Obstet Gynecol* 138:880–892, 1980

25. Curran JW: Economic consequences of pelvic inflammatory disease in the United States. *Am J Obstet Gynecol* 138:848–851, 1980

26. Mishell DR, Bell JH, Good RG, et al: The intrauterine device: A bacteriologic study of the endometrial cavity. *Am J Obstet Gynecol* 96:119, 1966

27. Burkman RT: Intrauterine device use and the risk of pelvic inflammatory disease. *Am J Obstet Gynecol* 138:861–863, 1980

28. Westrom L, Bengtsson LP, Mirdh PA: The risk of pelvic inflammatory disease in women using intrauterine contraceptive devices as compared to non users. *Lancet* 2:221, 1976

29. Burkman RT: The woman's health study association between intrauterine device and pelvic inflammatory disease. *Am J Obstet Gynecol* 57:269–276, 1981

30. Senanayake P, Kramer DG: Contraception and the etiology of pelvic inflammatory disease: New perspectives. *Am J Obstet Gynecol* 138:852–860, 1980

31. Huggins GR, Giuntoli RL: Oral contraceptives and neoplasia. *Fertil Steril* 32:1–23, 1979

32. Livosli VA, Stadel B, Kelsey JL, et al: Fibrocystic breast disease in oral contraceptive users: A histopathological evaluation of epithelial atypia. *N Engl J Med* 299:381, 1978

33. Armed Forces Institute of Pathology, Hepatic Branch and Center for Disease Control, Bureau of Epidemiology, Family Planning Evaluation Division: Increased risk of hepatocellular adenoma in women with long-term use of oral contraceptives. *Morbid Mortal Weekly Rep* 26:293, 1977

34. Merritt CG, Rosenberg SH, Edington B, et al: Age at first coitus and choice of contraceptive method: Preliminary report on a study of factors related to cervical neoplasia. *Socio biology* 22:255, 1975

35. Meisels A, Begin R, Schneider V: Dysplasias of the uterine cervix. Epidemiological aspects: Role of age at first coitus and use of oral contraceptives. *Cancer* 40:3076, 1977

36. Harris RWC, Brinton LA, Skegg DCG, et al: Characteristics of women with dysplasia or carcinoma in situ of the cervix uteri. *Br J Cancer* 42:359, 1980

37. Newhouse ML, Pearson RM, Fulherton JM, et al: A case control study of carcinoma of the ovary. *Br J Prevent Soc Med* 31:148, 1977

38. Tatum HJ: Clinical aspects of intrauterine contraception circumspection 1976. *Fertil Steril* 28:3, 1977

39. Israel, R, Shaw ST, Jr, Martin MA: Comparative quantitation of menstrual blood loss with the Lippes loop, Dalkon Shield and Copper-T intrauterine devices. *Contraception* 10:63, 1974

40. Pharriss BB: Clinical experience with intrauterine progesterone contraceptive system. *J Reproduct Med* 20:155, 1978

41. Cates WR, Schulz KF, Grimes DA, et al: The effect of delay and method choice on the risk of abortion morbidity. *Fam Plan Perspect* 9:266–272, 1977

42. Grimes DA, Cates WR: Abortion: Methods and complications, in Hafez ESE (ed): *Human Reproduction Conception and Contraception.* Hagerstown, Harper & Row, 1980, p 796

43. Tietze C: Repeat abortions: Why more? *Fam Plan Perspect* 10:286–288, 1978

44. Weinberger MH: The blood pressure effects of oral contraceptives, in Sciarra, JJ, Zatuchni, GI, Speidel, JS (eds): *Risks, Benefits and Controversies in Fertility Control.* Hagerstown, Harper & Row, 1978, p 230

45. Freeman EW, Rickles K: Adolescent contraceptive use: Current status of practice and research. *Am J Obstet Gynecol* 53:388–394, 1979

46. Edwards LE, Steinman EM, Arnold KA, et al: Adolescent pregnancy prevention services in high school clinics. *Fam Plan Perspect* 12:6–14, 1980

CHAPTER 7

Venereal Disease in Adolescents

Gregory C. Bolton

1. INTRODUCTION

Adolescent venereal disease currently is gaining acceptance as a significant topic of concern among public health officials, physicians, and even the public at large. The proper management of the adolescent patient with venereal disease requires consideration of the medical, epidemiologic, and behavioral aspects of venereal disease control.

The young female progresses from the initial appearance of secondary sex characteristics to early sexual maturity. The child's body undergoes complex changes necessary for adult sexual functioning, and the sex drive becomes awakened. Consequences of adolescent sexuality and sexual activity are the increases in teenage pregnancies and teenage sexually transmitted diseases. Premarital intercourse is beginning at younger ages and that the extent of premarital intercourse is probably on the increase.

With this increase in sexual activity by the young, it is not surprising to find that venereal disease statistics also have risen. The increase of *Neisseria gonorrhoeae* infection in the teenage group has risen much more rapidly than in the general population (one out of every four reported cases are adolescents age 10–19). Syphilis by contrast is less prevalent and only accounts for 0.5% of adolescent venereal diseases. It is conservatively estimated that in the United States about one adolescent actually becomes infected with either gonorrhea or syphilis each minute. Reasons for this upswing are multiple. First is the changing sexual mores of the teenage population, second, the earlier age of onset of sexual activity, and third, the greater range of sexual partners. A fourth possible

Gregory C. Bolton • Department of Obstetrics and Gynecology, Pennsylvania Hospital, Philadelphia, Pennsylvania 19107.

reason is the increase in reported case findings, and finally, the superficial knowledge that adolescents have of venereal disease.

Adolescents are particularly at high risk of venereal disease because they are sexually active and are often late in seeking medical attention. They are also less likely than older persons to form a monogomous relationship, and they are also less likely to take appropriate protective measures before engaging in sexual activity. Also, when they become infected, young persons may delay seeking the proper medical attention. This is secondary to their lack of accurate information concerning symptoms, consequences, and the appropriate treatment of sexually transmitted diseases. They also have a mistrust of medical personnel for fear their confidentiality could be violated. Also, the inadequate financial resources they possess often inhibit them from seeking private medical care. Because of the extensive venereal disease epidemic among adolescents, physicians whose practices include this age group should expect to see more venereal diseases.

Venereal diseases are epidemic in the United States alone. Venereal diseases are among the most prevalent and dangerous of our communicable disease states. The term *venereal diseases* (VD) seems to be loosing favor in the contemporary literature, and *sexually transmitted diseases* (STD) will be the designation in the 1980s. The sexually transmitted diseases are diseases of human behavior and are influenced greatly by socioeconomic, psychologic, and cultural factors. Some of the diseases are obviously more important than others because of their sequelae, but in aggregate they are the major cause for patient visits to physicians and can include some of the most common and important infectious diseases. Because reports limited to the sexually transmitted diseases of adolescent women are not abundant in the literature, much of the following review will depend on studies of adults. However, the ready application to the teenage patient is quite clear.

2. OVERVIEW OF SEXUALLY TRANSMITTED DISEASES

Of the venereal diseases originally considered important enough to report, only syphilis and gonorrhea are both frequent and serious enough to merit public attention. The others—chancroid, granuloma inguinale, and lymphogranuloma venerum—are now seen only rarely, and there appears to be general agreement that federally sponsored health efforts on these diseases is unwarranted. However, the number of new sexually transmitted diseases appears to be increasing and these new sexually transmitted diseases include trichomonaisis, herpes Types I and II, nongonococcal urethritis, nonspecific vaginitis, and genital warts. All of these are estimated to be more common than syphilis; however, the long- and short-range effects on society and individuals have not been completely studied. Specific disease characteristics of some of the more common sexually transmitted diseases are described.

3. GONORRHEA

The past decade has witnessed advances in the understanding of the etiology, pathogenesis, epidemiology, clinical manifestations, and management of gonorrhea. Gonorrhea is perhaps the most commonly reported communicable disease in the United States. Gonorrhea is classically defined as an inflammation of the mucosa of the genital–urinary tract caused by the gonococcus (*Neisseria gonorrhoeae*). Infection occurs either by hematogenous spread from a primary genital–urinary site or by direct innoculation resulting in a wide clinical spectrum of signs and symptoms. *N. gonorrhoeae* is a spherical gram-negative organism, which metabolizes glucose but not fructose or maltose. Having man as its only natural host, the organism is a known pathogen for the columnar and transitional epithelium of the genital–urinary tract. Involvement of the female genital tract is one of sequential infection due to continuous bacterial replication, dissemination along epithelial surfaces, and by submucosal lymphatic spread. This is in contrast to bacteria such as group A β-hemolytic streptococci, which penetrates the epithelium and pursues a transorgan route. The gonococcus has a tendancy to develop resistance to antibiotics, exemplified by the initial effectiveness of sulfonamides and the rapid development of resistance to them. Penicillin was developed near the end of World War II and became the new miracle drug of choice for all gonorrheal infections. Resistance to this drug has now occurred, and we now have complete resistance to some Southeast Asian strains and in a few North American isolates.

3.1. Epidemiology

The epidemic nature of gonorrhea is clearly recognized. Estimates, which are adjusted for underreporting by private physicians, show 1–3 million infections occurring yearly in the United States. The incidence of gonorrhea is highest in the young, poor, unmarried, and sexually active individuals. It is most common in sexually active subgroups, 5–20% of prostitutes in one study, while almost never reported in adult virgins. The age range peaks at 20–24 years of age followed by the 15- to 19-year-old group in which it is reported as high as 35% of the total number of cases. Gonorrhea tends to develop in the sexual partners of infected persons and rarely among other types of contacts. The disease usually begins with and is frequently limited to the sexual organs. The susceptible surface epitheliua are the urethra, endocervix, rectum, pharynx, and conjunctiva. In the male, an uncomplicated urethritis develops in 90–95% of cases and is distressing enough to result in early diagnosis and treatment. In women, the infection may go unrecognized because symptomatology is absent or nonspecific in up to 80% of cases.

The inability to eradicate widespread gonorrhea is due in part to the great reservoir of infection in asymptomatic males and females as well as the lack of development of self-immunity after an initial infection. Most cases are transmitted from asymptomatic individuals and from those who do not recognize the atypical signs and symptoms. As mentioned before, this unfortunately is the case with most teenagers. Although venereal in transmission and relatively susceptible to death by drying, the organism has been recovered after several hours from the surface of fomites to which a large innoculum had been applied. Although not proven as a natural method of transmission, this is of possible importance as a means of infection by nonvenereal transmission primarily in younger children.

3.2. Clinical Spectrum

The intimate contact of the gonococcus with epithelial or mucous-secreting cells appears to be of prime importance in initiating infection. Generally the initial symptoms due to infection with gonococcus appear 3–5 days after sexual exposure to the carrier. The risk of acquiring gonococcus after a single exposure has been estimated at 25–35%.

3.3. Sites of Gonorrheal Infection

The susceptible areas most exposed to infection in young women are the urethra and cervix. Symptoms produced are minimal and patients are frequently unaware of infection. Urinary frequency, slight dysuria, and leukorrhea may be part of what is called the urethral syndrome (dysuria with negative standard urine cultures). This is in contrast to the initial infection in heterosexual males where symptoms (dysuria and purulent discharge) occur in 95% of patients within 2 weeks of exposure.

3.3.1. Endocervicitis

Asymptomatic endocervicitis is the most common form of gonorrheal infection in females. The presenting symptom is vaginal discharge, classically described as a purulent one, but it may be scanty and mucoid in appearance. The gonorrheal discharge is described dogmatically in most textbooks as having a particular feature; however, descriptions vary with the only common factor being a distinct alteration of the normal discharge. The infection may also attack the pariurethral glands and lead to Bartholin's duct abscesses.

3.3.2. Prepubescent Vulvovaginitis

N. gonorrheae is reported to be the commonest cause of bacterial vulvovaginitis in the prepubescent female. Because the pediatric cervix has a cryptoform configuration and few endocervical glands, it is found to be less susceptible to infection. At the same time, the thinness of the vaginal wall, absence of estrogenic effect, and alkaline pH make the vagina a more susceptible site. Gonorrhea therefore in this age group is a vaginal infection rather than cervical one, is confined to the lower genital tract, and rarely spreads systemically. The most common symptoms, excoriation, edema, and erethema, often are accompanied by dysuria and itching. Attention then is diverted from the genital tract to the urinary tract. It is important for the physician to consider and identify gonorrhea infection in these girls: (a) to prevent further transmission within the family; (b) to identify such infections as possible markers of sexual abuse; (c) to forestall rare spread to other organs although acute pelvic inflammatory disease (PID) is rare during the preadolescent period. Secondary gonorrheal vulvovaginitis is epidemic in institutions for young women from infected bed clothes, towels, and bathtubs.[1] Infections in patients less than 1 year of age are attributed to poor parental hygiene. In ages 1–9 infection is usually secondary to a decrease in hygiene plus molestation. In ages over 9 infection is attributed to voluntary sexual contact.

3.3.3. PID

If the acute gonorrheal endocervicitis remains untreated, the infection may spread to involve the endometrium, fallopian tubes, ovaries, and pelvic peritoneum. This complex syndrome is the most frequently encountered, serious, and incapacitating complication of gonorrhea. Approximately 90% of women with acute disease complain of lower abdominal pain beginning with or increasing at the time of menses. This is in contrast to patients with nongonococcal inflammatory disease where the onset of symptoms is evenly distributed throughout the cycle. Vaginal discharge is a symptom in 55% of cases, menstrual irregularity occurs in 33% of cases, and dysuria is found in 10% of cases. The initial pelvic exam reveals adnexal tenderness in 90% of the cases, a palpable mass in 50% of the cases, and universally pain with motion of the cervix is noted. A temperature greater than 100.4°F is noted in about 40% of women. The gonococcus is cultured from the cervix in 80% of initial episodes of PID but drops to 50% with three or more recurrent infections. This suggests then that the initial episodes primarily are gonococcal and subsequent episodes involve other pathogens and are probably polymicrobial in origin. The intrauterine device (IUD) also predisposes females to both gonococcal and nongonococcal PID. These devices are known to increase the risk factor three times in users, four times in nulligravid users, and even higher for women who have the IUD in place for over 5 years.

Generally, initial symptoms appear 3–5 days after sexual exposure to the carrier. Examinations rarely are performed because symptoms are not severe, but the mucosa of the urethral meatus and orifices of pariurethral ducts are red and edematous. Pus usually can be expressed from the urethra and Skenes duct. The cervix is inflammed and green or yellow purulent discharge exudes from the external os. Gonorrheal endometritis is transitory and often heals spontaneously without leaving its mark. However, the irregular spotty bleeding that some women describe often occurs secondary to such acute superficial infection. The organisms spread quickly and bilaterally to the endosalpinx and involve the ovaries, cul de sac, and pelvic peritoneum by virtue of the pus that pours from the fimbriated ends of the tubes. This produces pelvic peritonitis. Symptomatically, there is fever, nausea and vomiting are not uncommon, and pain is moderate to severe and usually present in both quadrants of the lower abdomen. Physical examination reveals tenderness and rigidity in the lower abdomen, which may be accompanied by distention secondary to paralytic illeus. On pelvic examination, evidence of infection of the external genitalia and cervix may be present, and universal discomfort secondary to cervical motion often is noted along with tenderness in the lateral fornices. Laboratory tests can help to establish a diagnosis. The erythrocyte sedimentation rate and white blood cell count generally are elevated. Culdocentesis and laparoscopy are more accurate diagnostic tests than endocervical cultures and are indicated for patients with acute pelvic pain than cannot be differentiated from other pathologies, such as appendicitis, ectopic pregnancy, septic abortion, and rupture of an ovarian cyst. Without treatment symptoms usually subside in 7–10 days and are gone in 21 days. The gamut of gynecologic symptoms described as residual of gonorrhea include pain, abnormal uterine bleeding, and dyspareunia. Subsequent infections will be more serious because of the anatomically distorted pelvic structures with altered blood supply and altered ability to respond to infection.

3.3.4. Disseminated Gonorrheal Infections (DGI)

Occasionally gonorrhea enters the bloodstream from an isolated spot of localized infection. The systemic dissemination of *N. gonorrheae* (arthritis, dermatitis syndrome) occurs in 1–3% of patients harboring the organism and is the most frequent cause of infectious arthrtis and the most common cause of acute arthritis in young adults. DGI occurs in two stages. Initially gonococcal bacteremia and positive blood cultures are found in 50% of patients. Fever and chills are often associated with skin lesions of the extremities specifically the palms and soles. These begin as papules leading to petechial lesions that may progress eventually to vesicles and often become necrotic in appearance. The patient complains also of polyarthralgis, polyarthritis, and tenosynovitis. If untreated, the second stage is a septic joint. Blood cultures are often negative with no bacteremia and arthritis disappears. No new lesions appear, and the symptoms often localize to a single joint with the development of a purulent effusion often accompanied by destruction of the synovium.

3.3.5. Gonococcal Conjunctivitis

Gonorrhea in the adult usually reaches the eyes via the fingers and contact with the primary infection elsewhere. The disease manifests itself as an acute purulent conjunctivitis with copius discharge. Rapid involvement of the cornea, which may heal with scarring or may lead to frank perforation, is often the result. Reiter syndrome should be considered when accompanied by urethritis. Gonococcal ophthalmia neonatorum now is uncommon in the United States. Before the introduction of silver nitrate, gonorrhea was reputed to be responsible for 12% of all blindness. The consequences of delivery of an infant through an infected birth canal warrant the routine culturing of pregnant women at their initial visit and in certain high risk groups again during the last trimester. In addition, the instillation of 1% silver nitrate solution is appropriate at the time of delivery.

3.3.6. Gonococcal Perihepatitis (Fitz–Hugh–Curtis Syndrome)

Gonococcal PID may be accompanied by perihepatitis secondary to a direct extension from the pelvic focus. Infected patients often present with right upper-quadrant pain, hepatic tenderness, transient elevation in liver enzymes, friction rub, and often an abnormal oral cholecystogram is reported. There is an association of the disease with salpingitis in adolescents and should be included in the differential diagnosis of adolescents with salpingitis and upper abdominal tenderness.

3.4. Diagnosis of *N. gonorrheae*

The presence of intracellular gram-negative, kidney-shaped diplococci within polymorphonuclear cells obtained from an appropriate source is virtually pathognomonic of *N. gonorrheae*. The most reliable method to document gonococcal infection is demonstration of the organism by gram stain or culture. The major advance in the management of gonorrhea was the development of selective media. The modified Thayer Martin formula contains antibiotics to supress the overgrowth by proteus. An atmosphere rich in carbon dioxide is also needed for cultures inoculated away from the laboratory. Presumptive diagnosis is made when organisms growing on this selective medium with the typical morphology produce oxidase, but confirmation requires subjecting the organisms to sugar fermentation. Single cervical cultures are positive in 80–85% of the cases, and of all addition sites the biggest yield is obtained by culturing the anus and oral pharynx.

3.5. Therapy

The Centers's for Disease Control along with the public health service have formulated new recommendations for the treatment of gonorrhea.

1. Uncomplicated urethritis or cervicitis is treated adequately with aqueous procaine penicillin G. It should be administered as 4.8 $\times$ 10^6 units im divided into two doses and injected at different sites at one visit, along with 1 g probenicid po just before the injections. Patients in whom oral therapy is preferred may be given 3.5 g ampicillin together with 1 g probenicid administered at the same time. Patients who are allergic to penicillin or have had previous anaphylactic reactions may be treated initially with 1.5 g po tetracycline hydrochloride, followed by a 0.5 g dose qid for a total dosage of 9.5 g. Alternatively, patients allergic to penicillin may be treated with 2 g im spectinomycin hydrocloride in one injection.
2. Follow-up cultures should be obtained from the infected sites 3–7 days after completion of treatment. Tests of cure cultures should be obtained from the anal canal of all women who have been treated for gonorrhea.

When required by state legislation or indicated by local epidemiologic considerations, effective and acceptable programs for prophylaxis of neonatal gynecology ophthalmia include ophthalmic ointment or drops containing tetracycline or erythromycin or a 1% silver nitrate solution instilled in the eyes at delivery.

Equally effective treatment schedules studied in small numbers of patients with arthritis–dermatitis syndrome include the following: (a) 3.5 g ampicillin with 1 g probenicid followed by 0.5 g ampicillin po qid for 7 days; (b) 0.5 g tetracycline po qid for 7 days; (c) 10 $\times$ 10^6 units/day aqueous procaine penicillin G until improvement occurs followed by 0.5 mg ampicillin gid to complete 7 days of antibiotic treatment.

Special consideration should be given to the possibility of coincident syphilis. All patients with the diagnosis of gonorrhea should have a serologic test for syphilis. Negative patients without clinical signs of syphilis are treated with the recommended schedule of aqueous procaine penicillin G and need no serologic follow-up since that regimen is effective against incubating syphilis. Patients treated with ampicillin, tetracycline, or spectinomycin should have follow-up serology 3 months after therapy to detect untreated syphilis. Patients who have coincident syphilis manifested clinically or serologically should receive additional therapy appropriate to the stage of syphilis.

Hospitalization should be considered for patients in whom the diagnosis is uncertain, if there is a suspicion of pelvic abscess, if the patient is pregnant, or if she cannot reliably take or retain medications. Hospitalized patients may be treated with 20×10^6 units/day aqueous penicillin G, iv until there is definite clinical improvement followed by 500 mg ampicillin qid to be taken for 10 days of therapy. Since it is not possible to differentiate gonococcal from nongonococcal or polymicrobic salpingitis, many physicians use an aminoglycoside and an antibiotic specifically effective against *Bacteroides fragilis* in addition to the above programs.

4. SYPHILIS

The total number of reported cases for all ages of primary and secondary syphilis is upward of 21,000 a year. Approximately 30% of these cases occur in females. Syphilis in the heterosexual adolescent population is said to account for only 0.5% of all venereal diseases. The responsible organism is the spirochete *Treponema pallidum*. After infection, the female is often asymptomatic and may unknowingly infect others. In addition, irradication of this disease is hampered by the failure of physicians to report cases of infection to local public health authorities for tracing of contacts.

Following an incubation period of 10–90 days, the patient develops a primary papule which progresses to become a solitary, indurated, nontender ulcer known as a chancre. Usually present on the genitalia, 5% are noted at other sites, such as lips, breasts, or mouth. If identified at this point, serologic tests may be negative, but diagnosis still can be made by utilizing dark-field examination of material from the lesion. A reagin test for syphilis (VDRL, RPR) should be performed. If the dark-field test is negative and the reagin test is negative, repeat tests should be performed on patients for at least 3 months before a diagnosis of syphilis can be excluded. In those patients with primary syphilis who go undiagnosed or inadequately treated, many will proceed to develop secondary syphilis. This is commonly manifested by systemic signs and symptoms—fever, myalgia, headache, loss of appetite, generalized maculopapular skin rash, mucus membrane lesions including patches and ulcers, condylomata lata, and generalized lymph adenopathy. Meningitis, hepatitis, nephritis, alopecia, or iritis may also occur. The most common manifestations are those of skin and mucosa, and the blood serologic tests are almost always positive. Latent syphilis is defined in a patient with no clinical or historical findings to suggest syphilis but with positive serologic evidence of the disease. Patients with latent (hidden)syphilis of indeterminate duration should be evaluated for potential asymptomatic neurosyphilis. This is accomplished by examination of the spinal fluid. Although latent syphilis is not an immediate concern to the adolescent. There is still some question as to whether adequate treatment of early syphilis always prevents neurosyphilitic complications. Despite the best therapy some persons will develop these late complications for reasons unknown. Perhaps this uncertainty should be made known to the adolescent sexual activist.

4.1. Pregnancy and Syphilis

Needless to say, all pregnant women and girls should have a serologic test for syphilis at the time of the first prenatal visit. In women suspected of being at risk for syphilis, this should always be repeated in the third trimester even if negative initially. Active syphilis in pregnancy, whether preexisting or acquired during pregnancy, is a serious disease with great potential hazard to the fetus. A practical clinical consideration should be emphasized in dealing with pregnant adolescents. They characteristically come late for prenatal care. Thus, the attending medical team must be aware that the time for diagnosis, treatment, and follow-up of pregnant syphilis patients is foreshortened. Although meaningful statistics are not readily available, the following seems to be an acceptable assessment of the course and prognosis of syphilis cases. If no treatment is given, approximately one-third of the persons affected with syphilis may undergo a spontaneous cure, about one-third will remain in a latent state, and about one-third will develop the serious consequences of latent syphilis.

4.2. Treatment

The Centers's for Disease Control have recommended the following for the treatment of syphilis.

4.2.1. Incubating Syphilis

Women exposed to syphilis, or at high risk (e.g., sexual assault victim or diagnosis of gonorrhea), should be treated as for syphilis of less than

1 year duration or with 2.4 $\times$ 10^6 units aqueous procaine penicillin G in each buttock plus 1 g probenicid po. Treatment of syphilis of less than 1 year duration consists of 2.4 $\times$ 10^6 units benzathine penicillin G im at one visit, or 600,000 units/day aqueous procaine penicillin G for 8 days for a total of 4.8 $\times$ 10^6 units is recommended. If penicillin allergy is present, give 500 mg tetracycline hydrochloride qid for 15 days or 500 mg erythromycin qid for 15 days. For syphilis of over 1 year or unknown duration, treat with 2.4 $\times$ 10^6 units benzathine penicillin G, im weekly for 3 weeks or 600,000 units/day aqueous procaine penicillin G, im for 15 days for a total of 9.0 $\times$ 10^6 units. If penicillin allergy is present, give tetracycline hydrochloride 500 mg erythromycin po qid for 30 days. For syphilis in pregnancy, if the patient is not allergic to penicillin treat, the same as in the nonpregnancy patient. If she is allergic to penicillin, avoid the use of tetracycline and use erythromycin in the same dosage as the nonpregnant patient.

4.2.2. Follow-Up

After treatment, patients with early syphilis should be seen at 3-month intervals for at least 1 year and a quantitative reagin test should be performed at each visit. The titer should decline in most patients so that little or no titer is present at the end of a year. All patients with a positive spinal fluid should have spinal fluid testing at least every 6 months for 3 years. In addition, a spinal fluid examination should be done on those patients who wish to receive therapy other than the recommended penicillin regimen.

5. HERPES GENITAL INFECTION

Genital herpes infection is caused by a virus called herpes simplex. There are two strains of this virus, Type I and Type II, which may cause disease in humans. The majority of oral infections are caused by Type I strain, whereas the majority of genital lesions are caused by the Type II strain. Either strain, however, can cause disease on either the mouth or the genitals. Herpes genital infection is a sexually transmitted disease passed along by intimate physical contact with an infected person. In addition, the disease can spread by direct contact, with an infected area transmitting the virus to another site of the body.

Herpes genital infection is not a reportable disease in the United States. However, scientists estimate the incidence of herpes infection by sampling several clinics and have found that approximately one case of herpes infection is diagnosed for every ten cases of gonorrhea. Since there may have been as many as 2.7 million total cases of gonorrhea each year, the estimated number of new cases of gonorrhea each year, the estimated number of new cases of herpes is approximately 300,000 per year. It is also estimated that 9–35% of the general population have been exposed at some time to the virus. Herpes genital infection occurs in 1.6–8% of all patients seen in venereal disease clinics and in 0.2–2% of patients seen in gynecologists private offices.

In less than 1% of asymptomatic pregnant women, the virus can be cultured from the cervix. The sores of herpes virus infection generally develop 3–7 days after infection, often beginning with burning or tingling. Usually these are raised sores or fluid-filled blisters, which either spontaneously resolve or rupture to form shallow painful sores, which then scab and heal. The lymph nodes (glands) in the groin may swell and become tender. The initial infection lasts from 14 to 28 days. Women tend to have more discomfort than men. Infection of the cervix or vagina, however, does not always cause symptoms. Even though the sores go away, the virus frequently remains in the nerve tissue of the body and possibly in the skin. The virus can multiply at a later date and cause sores again. The number of recurrences is very unpredictable. Several stimuli such as fever, sunlight, menstruation, and emotional stress tend to trigger recurrences in some patients. Many patients can predict when sores are about to recur by noticing early symptoms of tingling, burning, or itching where the sores eventually arise. The sores of recurrent infection generally last for about 7–14 days.

Currently we do not have accurate information to determine the rate of spread of the disease from one individual to another. Generally the infection is spread when the individual has sores. During the dormant period between recurrences, the individual is less likely to transmit the disease. We encourage individuals to avoid sexual intercourse during the time they might have sores. If both partners are already infected, this advice may not apply. Condoms will decrease the chance of transmission.

5.1. Diagnosis

Physicians generally diagnose herpes genital infection on the basis of the patients history and examination of the sores. A blood test to determine levels of antibody against herpes may be done, but it does not reliably distinguish between either initial or recurrent infection or between Type I and Type II infection. A Pap smear of the lesion will help identify the virus on cytologic preparation. The most definitive means of diagnosis is culture; however, this technique is not available in all communities. Studies have suggested that the herpes genital infection predisposes women to develop cervical cancer. These studies are not conclusive; however, because both diseases have similar predisposing factors, both occur in young, sexually active groups, generally in women who have early frequent intercourse, and often in women who have multiple partners. The following data support this relationship: More antibodies to herpes simplex virus have been found in cervical cancer patients than in control patients, indicating a five fold increased risk; herpes simplex antigens have been found in cervical cancer cells; and women who develop cancer of the cervix more often give a history of herpes infection than do control patients. Women who have had genital herpes infection should have a cervical cancer check (Pap smear) at least once every 6 months. Early detection treatment can cure cervical cancer.

5.2. Neonatal Infection

Women who have herpes genital infection have as much as a threefold higher rate of spontaneous miscarriage. Herpes infection increases the risk for premature delivery. Herpes simplex virus can also cause severe disease in infected infants. If a mother is infected at 32 weeks' gestation or later, there is an approximate 10–20% chance that her infant will be infected; this risk is greatest if the infant is exposed to active infection in the mothers birth canal during delivery. Infants can also be infected while still inside the womb, although such infections are rare. If delivered through an infected birth canal, the baby has a 40–60% chance of being infected. The mortality rate of herpes of the bloodstream of an infected baby may be as high as 50%. Pregnant women who have had a history of herpes infection or who develop this infection during pregnancy should notify their obstetrical service and precautions can be taken.

5.3. Therapy

No well-documented, sage, 100% effective treatment exists. Many different agents have been tried in therapeutic studies; however, the definitive means of therapy has not been determined. Most clinicians recommend soaks to keep the area clean, drying agents to hasten healing of the lesions, and analgesics when needed.

6. CONDYLOMA ACUMINATUM

Condyloma acuminata (genital warts) are known to be transmitted through sexual intercourse. The causative organism is a papilloma virus. The risk of developing the disease following sexual activity is greater than

50% even though it may take 2–3 months to develop. The individual lesions are fleshy, soft, moist excrescences and are usually located on the perineum, vagina, cervix, and penis. The lesions can be few in number or coalesce to such a large area as to cause problems with urination, defecation, and even vaginal delivery. The diagnosis is made by clinical observation, but biopsy proof is valid in cases when differentiation of condyloma lata is considered. The close association with these lesions and other vaginal infections (i.e., trichomoniasis and moniliasis) is well documented. Treatment depends on the size and location of the lesions as well as the pregnant and nonpregnant status of the patient. Therapy with 20% podophyllin solution (mixed with a benzoin base) applied locally is adequate for small lesions. Applications can be repeated weekly, never intravaginally (anaphylaxis) and never during pregnancy. If necessary, surgery, cryosurgery, or electrocautery can be used to rid the patient of this disease.

7. PEDICULOSIS PUBIS (CRAB LICE)

Crab lice (*Phthirus pubis*) are parasites which are transmitted through sexual intercourse as well as shared contact with clothes or linen. The nits or eggs are deposited at the base of hair shafts and after 7–9 days, they hatch and the new lice attach themselves to the skin of the host. An erythematous papule may result from the small bites, and secondary infection can follow. The diagnosis is made by the history of severe itching or the actual sighting of the lice moving on the skin surface. Positive identification can be made with the aid of a magnifying glass or a microscope. Treatment with Kwell® (1% j-benzene hexachloride) cream or shampoo is still the standard treatment. Treatment should be repeated in 4–7 days as well as concomitant treatment of sexual partners.

8. SCABIES

Scabies is a local skin infection caused by the female itch mite (*Sarcoptes scabies*). Transmission is again through sexual contact or close skin-to-skin contact. The ova are deposited beneath the skin by the itch mite and soon intractable itching and skin excoriation begin. Poor hygiene and living conditions are associated with the presence of scabies. Diagnosis is made by observation of the distinctive papular skin eruptions on the pubis, axilla, and flexer surfaces of the extremeties. There also may be characteristic burrows under the surface of the skin. Again, Kwell is the treatment of choice and should be applied to the entire body from the neck down and repeated in 24 hr. Infected bedding and clothing should be sterilized, and all sexual contacts treated as well.

9. CHANCROID

Chancroid is of the classic but rare venereal diseases in the continental United States. The causative agent is *Hemophilus ducreyi*, a coccobacilis that is nonmotile, nonacid fast, and gram-negative in characteristic. The infection may occur in conjunction with other genital infections, particularly genital herpes and syphilis. The classic lesion is a ragged, tender ulcer that is not indurated (soft chancre). The base of the lesion is covered with a gray or yellow necrotic exudate. There can also be unilateral tender inguinal lymphadenopathy associated with the skin lesions. The diagnosis is made by clinica appearance and exclusion of syphilis. A gram stain of the exudate reveals short gram-negative rods, and cultures may be performed on blood agar or media with blood derivatives. Complications of this disease include chronic fistulization and gland masses in the groin. Treatment is best accomplished with 1 g sulfasoxasole orally qid and 500 mg tetracycline qid for 10–14 days. Fluctuating gland masses call for aspiration.

10. GRANULOMA INGUINALE

Though fairly common in a few underdeveloped nations, this classic venereal disease has declined from a high of 2611 cases in 1949 to 75 cases in 1977 in the continental United States. It is more common among men than women and more common in the Southern states. Clinically it presents as single or multiple subcutaneous nodules, which may erode through the skin producing clean granulomatous, beefy red, painless lesions. Cytologically, intracytoplasmic rods (Donovan bodies) are found in large mononuclear cells. This is noted in biopsied Giemsa- or Wright-stained materials. The long-term consequences are rare but include elephantiasis, rectal strictures, and ulcerative and fistular lesions of the urethra, penis, and other genital structures. Treatment is primarily through using 500 mg tetracycline po qid for 2–3 weeks or 40 mg gentamicin im bid for 2 weeks.

11. *Chlamydia trachomatus*

An increasing number of sexually transmitted infections has been attributed to *Chlamydia trachomatis*. This pathogen has been known to cause trachoma, inclusion conjunctivitis, and lymphogranuloma venerum (LG). Because (*Chlamydia*) are obligate intracellular microorganisms that grow only within host cell cytoplasm, tissue culture methods similar to those used to recover viruses are required. (*Chlamydia*) attach only to collumnar or transitional epithelial cells, and with the exception of the serotypes L-1 to L-3 (LG), they usually do not invade deeply into tissue. Thus, most infections produced by severe types A through K have non-specific symptoms, such as a discharge, swelling, erythemia, and pain. Widespread or systemic symptoms are unusual except in the instances of superficial infections of tubal epithelium associated with PID, which can produce generalized peritonitis by contiguous spread, or in infections with LG serotypes, which often cause deep tissue envasion. *C. trachomatis* can be acquired at birth by contact with the infected cervicovaginal secretions and also during sexual contact with infected secretions. In one study, from 3 to 5% of unselected women had *C. trachomatis* isolated from the cervices. In a selected population, i.e., veneral disease clinics, the incidence of *C. trachomatis* is as high as 16%, whereas gonorrhea is estimated at 5% and herpes also at 5%. Cervical isolation rates of 4–13% have also been reported in many pregnant women in certain socioeconomic groups. The cervicitis associated with *Chlamydia* is accompanied in 30% of them by hypertrophic erosion of the cervix as well as in 40% a mucopurulent endocervical discharge.

The cervix is often friable and bleeds easily to touch. The mucopurulent cervicitis is either secondary to gonorrhea or *Chlamydia* until proven otherwise and thus requires antimicrobular therapy and treatment. Since *C. trachomatis* infections require specific indentification via isolation and tissue cell cultures, it is apparent that because such cultures are not universally available treatment should be with tetracycline for any nongonococcal mucopurulent cervicitis. The abnormalities on the surface epithelium may also cause cellular changes that lead to abnormal pap smears. The acute PID often attributed solely to gonorrhea can also be secondary to the *Chlamydia* organism. In Sweden, studies have shown that the incidence of gonococcal-induced PID is only as much as 17%, whereas 35% had been associated with *C. trachomatis*. A recent United States study out of Seattle has shown that the incidence of acute PID in the initial phase has been associated 20% of the time with *C. trachomatis*.

11.1. Inclusion Conjunctivitis

Approximately 40–50% of infants born through the cervix of patients harboring *C. trachomatus* have been found to contact inclusion conjunctivitis. This is noted to be the most common cause of neonatal eye infection. Silver nitrate therapy is of no value in a disease that begins usually 5 days after birth with the sudden onset of conjunctival hyperemia, and rapid progression of a mucopurulent discharge is best treated with tetracycline opthalmic eye drops.

11.2. Neonatal Pneumonia

Realizing that anywhere from 4 to 13% of pregnant women have *C. trachomatis* in the os at term, it is not surprising to find that approximately 40–50% of the babies born to these infected mothers develop *Chlamydia* inclusion conjunctivitis. Most neonates with the conjunctivitis also had pharyngeal colonization and described the syndrome of a rather distinctive type of interstitial pneumonia among infants of less than 6 months of age. The pneumonia had a gradual onset after several weeks, usually developing approximately 6 weeks after birth and was characterized by a rather repetitive stacatto-type cough. They were usually afebrile; however, other symptoms included congestion and conjunctivitis as well as bilateral interstitial infiltrates with hyperinflation of the lungs present on X-ray examination.

11.3. Diagnosis

The most sensitive method of diagnosis particularly from a genital site is to culture the organism on radiated McCoy or Hela cells. In the case of PID, the most desirable method is to have a direct culture from the fallopian tube. Even if the pus is sterile, Giemsa staining can show the presence of the organism. The presence is noted because of the visualization of multinuclear cells intracytoplasmic inclusions and cytomegaly. The Frei test, which in the past was the classic means of diagnosing this infection by using an indirect skin test, has now gone by the wayside, and the use of more specific and sensitive tests are now available.

11.4. Treatment

Attempts have been made to use the aminoglycosides in the treatment of *Chlamydia* infections. However, they do not inhibit the growth of the *Chlamydia* organism. Spectinomycin as well as ampicillin may have a minimal effect on the infectious process. Tetracycline, however, does remain the drug of choice for chlamydial type of infections.

12. VULVOVAGINITIS

Probably the most common infections seen in the office are still the female vulvovaginitises, *Candida albicans, Trichomonas vaginalis*, and nonspecific vaginitis. Because of the upsurge in interest in vulvovaginal diseases in recent years, today there is no longer a place for broad-spectrum shotgun therapy with these infections. Today because of the increased utilization of wet smears and cultures, a physician is often able to pinpoint the exact infection, and specific treatments can usually be prescribed.

12.1. *Candida albicans*

This infection is usually a flare-up of the endogonous organisms found in the rectum or vaginal area that cause a rather severe vulvitis rather than vaginitis. The vulvovaginitis secondary to infection with *C. albicans* is probably the most common cause of vulvar pruritis. The probable causes for increasing numbers in this infection in the past two decades include widespread local steroid application, broad-spectrum antibiotics, and possibly oral contraceptive usage. Sexual transmission is common, and it may well be that many female infections are secondary to sexual intercourse. The chief symptom of pruritis is caused by an allergic reaction to the fungal products or possibly the fact that these products might be toxins themselves. The chief sign is vulvar erythema limited to the vestibule, and often labial edema accompanies it. In chronic cases, fissuring and traumatic excoriation occurs, and these shallow erosions are occasionally confused with herpetic lesions. The diagnosis is relatively easy by means of a wet smear using 10% KOH solution or culturing the vaginal discharge on Nickerson slants. The treatment still remains relatively standard by using miconazol and mazole treatment schedules of 7 days' duration. More recent studies have shown equal success with clotrimazole bid on a 3-day treatment schedule.

12.2. *Trichomonas vaginalis*

T. vaginalis exhibits a wider spectrum of involvement than do other vaginal organisms. The disease is characterized by remission and exacerbation with spontaneous culture cures being very rare. The presence of a vaginal discharge is the foremost symptom of trichomonaisis classically described as a copious, frothy, greenish-tinged, and malodorous discharge. In the acute infection, the edema and inflammation of the vaginal papillae cause the well-known strawberry cervix and vagina. This disease is easily identified because of its intense pruritis and the highly mobile trichomonad visualized via microscopic examination of a physiologic saline wet smear. The zeal and experience of the observer in reading a wet smear often determines the case finding ratio. The patient may be cured and reinfected in rapid order unless her partner or partners are also treated. The treatment drug of choice remains metronidazol or Flagyl.® Not long ago, the treatment schedule was shortened from 250 mg three times a day for 10 ten days to 250 mg three times a day for 7 days. Several current studies have shown that a single stat dose (2 g) is equally as effective as a 7 day schedule. The other benefit that is gained by utilizing this schedule is the increase in patient and partner compliance. The incidence of side effects is no greater and most patients prefer the stat treatment dosage. In addition to nausea and the metallic taste, one must not forget the increased anticoagulant affect on Coumadin®, the reversible neutropenia, and the absolute contraindication of usage during the first trimester of pregnancy.

12.3. Nonspecific Vaginitis

The question of the entity of nonspecific vaginitis really remains unanswered, and one wonders if the answer really exists. It is accepted that in women presenting with a vaginal discharge and in whom a diagnosis of gonorrhea, trichomonaisis or candidiasis is made can be treated promptly and effectively. Those without such a microbiologic diagnosis and with so called nonspecific vaginitis have largely gone untreated or treated in a shotgun-like manner. Clinically *Hemophilis vaginalis* appears to be causally linked with the asymptomatic mild nonspecific vaginitis. *Hemophilus vaginalis* satisfies all Koch's postulates and is more common than *Trichomonas* as a cause for leukorrhea. Evidence linking *H. vaginalis* to the entity of nonspecific vaginitis is the fact that *H. vaginalis* is recovered more often and in higher quantities from women with nonspecific vaginitis than those with other forms of vaginal secretions or from normal women. Also, the improvement of clinical signs of the infection of nonspecific vaginitis correlates well with the eradication of the organism. Evidence for sexual transmission is based on the recurrence of *H. vaginalis*-associated nonspecific vaginitis after initially successful treatments following sexual reexposure. Also, the urethral cultures in 90% of the male partners of women with *H. vaginalis* revealed the predominant organism to be *H. vaginalis*.

12.4. Diagnosis

In a patient with symptoms and signs of abnormal discharge, the diagnosis of nonspecific vaginitis can be made with certainty by the following: First, the exclusion of candidiasis, trichomonaisis, and cervicitis must be made. Second, the demonstration or the presence of clue cells by microscopic examination of vaginal secretions diluted in a 1:1 normal saline wet mount examination. Clue cells are vaginal epithelial cells coated with coccobacilli forms of *H. vaginalis* to the extent that the borders of the cells are often completely obscured.

12.5. Treatment

The treatment of *H. vaginalis* generally has been frustrating. Sulfonamide-containing vaginal creams are ineffective perhaps because sulfonamides are uniformly inactive against *H. vaginalis in vitro*. Tetracycline therapy is also usually ineffective. Ampicillin, 500 mg, po qid for 7 days has been effective in most cases of nonspecific vaginitis. However, recent data support the fact that the most consistently effective therapy consists of 500 mg metronidazol bid for 7 days, which irradicates the organism and clears the symptoms.

13. CONCLUSION

It is now obvious that we are in the midst of an epidemic of sexually transmitted diseases. Noting that they are striking every social and ethnic group within our society and exacting from them their heaviest toll from patients age 15–30. Over the past few years, both gynecologists and pediatricians have become increasingly aware that the gynecologic needs of females in the pediatric and adolescent age group requires special attention. Our purpose was to review adolescent venereal disease in its clinical presentation, diagnosis, and current forms of management. This does not imply that the problem encountered by the adolescent age group are unique. In fact, if one were to list the common gynecologic complaints in females under the age of 18 it would be obvious that with few exceptions they encompass the same spectrum of conditions seen in the adult. Unfortunately, pediatricians have little or no formal training in gynecology and often approach the problem with insecurity. Gynecologists as well have little contact with this particular age group. The real need then is to develop an approach that eliminates fear, facilitates a comfortable yet thorough examination, and creates a relationship that allows for education and counseling as well at treatment for the youngster and his or her partner. Traditional avoidance prevents dealing with the total health care of our young patients. How we approach our adolescent patients with this type of problem often affects their ideas on lifetime health care as an adult.

BIBLIOGRAPHY

Branch G, Paxton RA: Study of gonococcal infections among infants and children. *Public Health Rep* **80**:347, 1965 *Social and Behavioral Aspects of the Sexually Transmitted Diseases, Sexuality Today and Tommorow.* 1976, p 134

Darrow WW: Venereal infections in three ethnic groups in Sacramento. *Am. J. Public Health* **66**:446, 1976.

Evans TN: Sexually transmitted diseases. *Am J Obstet Gynecol* **125**:116, 1976

Holmes, KK: Sexually transmitted diseases, in *Harrison's Principles of Internal Medicine.* New York, McGraw–Hill, 1980, Chap 113

Jerome E, Fleming, B, Miller, M: Gonorrhea at a teenage medical service. *Minn Med* **57**:245, 1974

Mumford DM, Smith PB: *Adolescent Pregnancy* (Perspectives for the Health Professional). Boston, Hall, 1980

Rigg CA: Venereal disease in adolescents. *Practitioner* **214**:199, 1975

Ryan GM: *Ambulatory Care in Obstetrics and Gynecology.* New York, Grune &Stratton, 1980, Chap 17

Stern MS, McKenzie RE: Venereal disease in adolescents. *Med Clin N Am* **59**:1395, 1975

Sparling P: *Current Problems in Sexually Transmitted Diseases* Chicago, Year Book Med Publ, 1979

U.S. Department of Health and Human Services: *Recommended Treatment Schedules for Syphilis.* Atlanta, Ga, Centers for Disease Control, 1981

U.S. Department of Health and Human Services: *Gonorrhea: Recommended Treatment Schedules.* Atlanta, Ga, Centers for Disease Control, 1981

CHAPTER 8

Rape of the Adolescent

Joseph A. Zeccardi

1. INTRODUCTION

Rape is sex without consent. It may take the form of either an assault or sex with one who is legally unable to consent. Rape is legally considered to be carnal knowledge by force and against the victim's will.[1] The patterns of rape vary according to age with the term *sexual abuse* being used more frequently for instances where a child unable to consent is enticed or has sex with older individuals. Sexual assault, whether it takes the form of a violent attack or sexual abuse of a minor, concerns the adolescent either directly or indirectly. As adolescence approaches, the incidence of violent attack by strangers increases significantly as does the frequency of sexual abuse where the assailant is known to the victim.[2] Adolescence also is the period during which the child must resolve the emotional effects of incest.

Studies done over the past decade delineate what is felt to be both an increased frequency of reporting and an absolute increased frequency of incest.[2,3] Generally, the findings have indicated that most of the victims were females with a small percentage of males. The assailant tends to be known by the child and is most frequently a member of the household. Of all cases of rape, approximately half involve children and adolescents. Assailants tend to be male and related to the victim, and the majority of the victims have no genital injury or laboratory findings.[4-9]

The clinician's role in rape includes the traditional role of the physician in obtaining an appropriate history and physical in order to properly treat the consequences of this condition both medically and psychiatrically. However, in addition to this traditional role, it is necessary for the physician to be sensitive to legal issues since the examination must also

Joseph A. Zeccardi • Department of Emergency Medicine, Division of the Department of Surgery, Thomas Jefferson University Hospital, Philadelphia, Pennsylvania 19107.

yield evidentiary material. Because of this legal issue, it is advisable that clinicians involved in collection of evidentiary material plan with the legal authorities in their area for appropriate evidence gathering. The method of charting and the particular material that is used for evidence vary from state to state and from community to community. In addition to the legal system, it is also important to work with volunteer and therapy groups who may exist in the area for the support of the sexual assault victim.

The medical examination of the victim includes the care of the physical and emotional needs of the victim and gathering of historical and factual data that will be helpful in legal proceedings. In addition to the proper medical care of injuries, prevention and treatment of possible venereal disease, pregnancy, and adequate medical follow-up with appropriate counseling to reduce emotional trauma are necessary.

The physician and medical staff are not to be judgmental. Patients do not ask to be raped by trying to look appealing, and the important difference is the consenting ability of individuals who agree to have sexual intercourse as opposed to one person violently forcing himself on another.

2. NOTIFICATION OF AUTHORITIES

Rape is a crime, and the law may require that physicians report any injury inflicted in violation of the penal code. Since one cannot know conclusively if an injury is a result of a penal code violation, a reasonable belief would necessitate reporting. This would indicate reporting, therefore, when the patient claims sexual assault or when the physician has reason to assume assault has occurred. In addition to state laws, many municipalities have ordinances requiring physicians to report sexual assault to a multidisciplinary group which manages these complaints. In children and adolescents, the additional reporting to child welfare agencies is required, especially, if the assault occurred within the household.

3. PERMISSION TO EXAMINE AND TREAT

Adolescents are usually under the legal age of consent. One needs to consider the various statutes that allow patients to consent for examination and treatment on their own. The reader is referred to Chapter 11, which deals with the legal rights of adolescents. In general, the emancipated minor is variously defined by individual states, and this may be related to their marital status or whether they have been pregnant. Some states would allow medical and health services to be provided to a minor to treat pregnancy or venereal disease. At any time when attempt to secure consent would delay treatment and increase the risk to the minor's life or health, consent of the parent or guardian for medical services is not legally necessary. In most other cases, it is advisable to obtain parental consent since successful long-range treatment of the adolescent depends on these supports.

An additional consent is necessary in order to share the information obtained through the examination and treatment with legal authorities such that prosecution may proceed. The patient may refuse to press charges and to share that information with the legal system. All information obtained should be retained by the hospital and may be released upon the written consent of the patient or guardian or may be subpoenaed. Without specific consent or court order, the only information released should be the report to the police of the injury and the patient's name and address in addition to the child welfare reporting. Some states have a right-to-know law, which exempts the state from giving to the public information that may be detrimental to the reputation or safety of the individual, thus, allowing the hospital to withhold the patient's name from the news media. Other consents will be necessary for the dispensing of diethylstilbestrol (DES) as emergency treatment against pregnancy after informing the patient of the possible side effects and dangers of estrogen therapy. A special permit is advisable.

4. THE POLICE INTERVIEW

In the physician's office or emergency room, the initial contact with the police ought to be conducted such that it occurs in a private setting and is preceded by a brief explanation by the clinical staff to assure the patient of the appropriateness of the interrogation. Some may feel that discussion with the police should not occur once the patient has disrobed, however, because of the acute nature of police work and the unavoidable delays that occur in managing a complex abuse case; it is sometimes necessary to support the police in their gathering of information. All attempts should be made to assure that the patient is comfortable with the interview and does not feel unduly exposed and vulnerable. It is not necessary nor appropriate for police to be present during the physical exam.

5. MEDICAL EXAMINATION AND TREATMENT

The physician responsible for the examination and treatment of the rape victim would best be prepared by not only reviewing the mechanics of the history and physical but also reviewing one's attitude about abuse and rape. There are many presumptions that are false about sexual abuse and rape, which tend to serve the purpose of making one comfortable with the issue but in fact, because of their inaccuracy, claim a significant part of the objectivity necessary to function appropriately in this realm. Society is full of misinformation about the nature of sexual abuse, its victims, and perpetrators. Readers are directed toward a publication of the United States Department of Health and Human Services entitled "Sexual Abuse of Children—Selected Readings."

The history itself should include a gynecological history appropriate for age, establishing the presence or absence of menarche; the activity postassault, such as changes of clothing, douching, and bathing; and the time of the last reported assault. Cases of child sexual abuse that are intrafamilial as opposed to assaults by a stranger tend to be reported late. Pertinent medical history should include detailed menstrual history, history of contraceptive use, and venereal disease history as well as the last coitus before or after the assault that may have been consensual. Ascertain the interests of the patient concerning hormonal pregnancy prevention, intrauterine device (IUD) insertion, abortion, menstrual extraction, or carrying a potential pregnancy to term. The rest of the history should concentrate on gathering information that would direct the physical, looking for other medical consequences of the assault or physical evidence of the assault, such as did restraining at the wrists occur such that marks might be noted. The child should be questioned carefully about the events using terminology that is familiar to the patient.

It is important to treat emergency conditions first; stop bleeding and assess for signs of penetration into the abdominal cavity. A general examination should be performed documenting the general physical appearance and demeanor of the patient. Usually, patients after a violent assault are in a quiet state. One should carefully assess for the presence or absence of marks on the clothing in addition to looking for rips or potential sites on material that may be either blood or semen. The body should be inspected for either the presence or absence of marks of violence with particular attention to any sites that were alleged to be traumatized. For example, petechia may be found in the fundoscopic examination if neck pressure impeded venous return just as bruising may occur in the area of direct neck pressure. It is important to remember that bruises and ecchymosis evolve over a period of time and may show as red marks early, which may later evolve into a more discrete ecchymosis. The presence of tenderness in an area that has been assaulted, although medically unimportant, is important legally since the physician's ability to assess tenderness and ascribe it to soft tissue injury becomes significant evidence in court. Any areas of the body that show dried blood ought to be scraped and those scrapings put into a dry paper envelope for testing in the laboratory. This should be marked "apparent area of dried blood" with specific reference to the site of collections. The same should be done for areas that appear to be dried semen. If complaints of oral or anal penetration are offered, appropriate studies should be taken and an examination done. The victim should be explicitly questioned as to where genital contact had been made.

A vaginal examination is necessary for determination of injury. Injury to the point of a need for medical intervention is rare. However, some evidence of minor injury may be found, which will be helpful in processing the case. A nonlubricated but water-moistened speculum should be used. The smallest speculum appropriate for any age victim should be utilized, since tissues are particularly sensitive postassault. A pelvic examination should include a general inspection of the external genitalia looking for evidence of trauma. Obvious foreign hairs that are noted should be collected in a paper envelope and labeled "apparent foreign hair." In the general inspection of the vulvar area, one should look for small ecchymoses as evidence of digital pressure. Examination then should proceed toward including a description of the orifice of the vagina.

The hymen, if it is intact, should be so described although it must be remembered that even in the presence of an intact hymen vaginal penetration may have occurred. (Another point to bear in mind is that, at least in Pennsylvania, the definition of penetration means between the labia as opposed to through the introitus.) The hymenal ring may show evidence of recent disruption, such as a mild reddening, small lacerations in the hymenal tissue, or obvious laceration with significant bleeding. Significant bleeding is rare. It must be remembered that many adolescents do not have identifiable hymenal tissue that would be disrupted by vaginal intercourse. The vaginal wall should be inspected for evidence of trauma, and the cervix should be visualized. Lacerations, contusions, and abrasions may be noted in the vaginal vault.

A bimanual exam should be performed, and appropriate studies should be obtained from the vaginal vault. A bimanual exam is performed to determine the size of the uterus and ovaries and to assess for trauma or the possibility of an already existing uterine pregnancy.

The physical examination should not become another assault. A speculum examination is not always necessary, appropriate, or indicated. The physician may collect specimens by obtaining either blind swabs or saline washes. This technique requires instilling approximately 5–10 ml of normal saline and aspirating the material. The material may be used for the same studies as outlined below. Alternative ways of examining the patient, such as the above-mentioned saline washes, and the alternative positions, such as a knee–chest or lateral position, may be more acceptable.

6. LABORATORY EVALUATION

In obtaining laboratory specimens, the material used for evidence must be handled carefully. It is important that all containers and slides be marked indelibly with the name, date, and some identifier of the patient, such as the chart number. An ultraviolet lamp may be used prior to collecting evidence. A Woods' lamp will cause semen to fluoresce and, therefore, indicate areas of the body or clothing from which evidence might be taken. The Woods' lamp is best used in the dark after visual adaptation. All specimens obtained should be labeled as to the site. Physicians should initial all specimens indicating that the chain of evidence goes from the patient to the physician and from the physician to the receiving person in the police department.

At Thomas Jefferson University Hospital, the following laboratory tests are obtained if the assault has occurred within 14 days of the examination. If it has been over 14 days, specimens for sperm, acid phosphatase, and blood group substances are not done. Although this 14-day period is at the extreme end of the biological curve for the finding of sperm, it is felt that there is still a likelihood of finding objective data and specimens are collected at that time.[10,11] Motile sperm have been reported to persist in the vagina from 3 to 24 hr and in the cervix from 110 hr to 7 days. Nonmotile sperm have been found in the cervix for 12 days and in the vagina for lengths of time varying from 14 hr to 14 days.[11,12] Specimens for the evaluation of venereal disease and pregnancy plus physical examination must be performed at any point that sexual assault is reported.

7. SPECIFIC LABORATORY TESTS

Smears for sperm are taken from the labia and posterior vaginal vault, cervix, and other areas of involvement. These may be studied either looking for motile sperm or, and more practically, looking for evidence of sperm on a dry glass slide. The variability of finding sperm is significant but is well worth investigating up to the 14th day.[10-12] Acid phosphatase is obtained from the same sites using a cotton-tip applicator, which is usually dropped into saline and utilized for the enzyme determination in the laboratory. Blood group substances may also be analyzed from the same specimen. Blood group substances, ABO, and Rh factors are excreted by some of the population into their secretions. Therefore, blood group substances that are foreign to the patient may be used as evidentiary material either to rule in or rule out a potential assailant.

Venereal disease testing should include a baseline serologic test for syphilis and cultures for gonorrhea obtained from the cervical canal and posterior fornix, plus other cultures from the anal or oral cavity if indicated. Pregnancy testing is useful at times to determine if a preexisting pregnancy is present because of the problems with DES. Saline washes, as mentioned before, may be an alternative way of obtaining the same information listed above. Swabs from the anal or oral cavity may be used for the same studies. Saliva is usually collected from the patient to determine if the patient is a blood group secretor in addition to obtaining blood for the typing of the patient. Rectal swabs and smears should also be obtained if indicated. Obvious foreign material from under the fingernails should be scraped into a dry paper envelope and appropriately labeled "fingernail scrapings." Obvious foreign hairs are handled similarly, and reference hairs should either be plucked or cut from the pubic area and appropriately labeled "reference hair." The patient's clothing should be carefully scrutinized as should the abdomen, thighs, and vulva looking for seminal fluid, alien pubic hair, or dried blood. This should be collected in clean dry envelopes; possible dried fluid should be collected on saline-soaked swabs and placed in sterile tubes or flaked off into an envelope. Any questionable foreign matter on the skin should be placed in an envelope and labeled.

8. MEDICAL THERAPY

In the more violent cases of rape, any major or life-threatening injuries should receive the highest priority. Life-threatening injuries should be attended to, then the medical therapy consisting of the treatment of venereal disease and the treatment of pregnancy consequences. As to the treatment of venereal disease, the patients are treated as uncomplicated gonorrhea unless an obvious diagnosis is established at the time of examination.

The treatment regimen is that recommended by the Communicable Disease Center and includes the following choices: (a) 4.8×10^6 units of aqueous procaine penicillin G im preceded by 1 g of probenicid po. The incidence of reactions to the procaine in aqueous penicillin and the risk of sensitization leads us to recommend the alternative of 3.5 g of ampicillin with 1 g of probenicid po. Since this large number of tablets may be nauseating and lead to vomiting, the patient should be observed for a minimum of 45 min before discharging the patient to be sure that the medicine is not lost. If the patient is allergic to penicillin, 1.5 g tetracycline po. immediately followed by 500 mg qid for 4 days may be substituted for a single dose of 2 g spectinomycin im may be used. For smaller patients, the treatment would be 75,000 units/kg aqueous procaine penicillin G as a single dose with 25 mg/kg probenicid or 100 mg/kg ampicillin po preceded by 25 mg/kg probenicid. A dose of 50 mg/kg amoxicillin preceded by probenicid is an acceptable alternative. If the patient is allergic to penicillin, tetracycline may be used in doses of 25 mg/kg initially followed by 40–50 mg/kg/day qid for a period of 7 days.

Potential pregnancy may be dealt with in a number of ways. The patient's interest in dealing with potential pregnancy as a consequence of sexual assault must be ascertained as must the preferences of the family. A number of alternatives are available. DES may be offered to the patient in doses of 25 mg bid for 5 days. If offered within the first 72 hr following unprotected intercourse, it has a 99% effectiveness in preventing implantation. However, adenocarcinoma in the adolescent female offspring of women exposed to DES in the first trimester has been reported and necessitates carefully informed consent. The incidence of vaginal adenocarcinoma in exposed female offspring is between 1.4 per 1000 and 1.4 per 10,000.[13-16] Other options include IUD insertion, elective abortion, or carrying any potential pregnancy to term.

9. EVIDENCE HANDLING

Microscopic examination and laboratory assessment of all evidentiary samples are usually done by the police pathologist rather than hospital personnel, since hospital personnel may not be legally accepted as experts. If the patient agrees to turn over clothing to the police, garments should be turned over by removing them without cutting through existing holes. Seam lines should be followed. They should not be shaken out but rather placed on a clean sheet of paper, which then should be placed inside a paper envelope or bag and properly labeled with evidence tape. Evidence tape is tape specially formulated such that its removal destroys the integrity of the tape, and it is obvious that someone has been tampering with the evidence. If the patient's clothing is torn, bloody or soiled and it is not practical to turn it over immediately to the police, the patient should be advised to turn the clothing over as soon as possible as it may be helpful in the investigation. It is wise to make the torn parts of the clothing known to the police who will follow-up.

It is important that a "chain of evidence" be maintained. That is, a method to be established to insure that all persons who were responsible for keeping or handling the evidentiary material can be easily traced. Consequently, it is advisable that evidence be handled by as few personnel as possible. Also, a form should be attached to each item of evidence showing the date, time, name of the person receiving the item, and from whom they were received. Evidence should be kept in an appropriate locked container until picked up by the police.

10. RECORDS

It should be remembered that the determination of whether or not a rape has occurred is the responsibility of the court rather than of those treating the victim. The medical record should not reflect any conclusions regarding whether a crime has occurred. Final diagnosis should be sexual assault complaint rather than sexual assault. Physician court appearances are reduced significantly if the medical record is properly kept and legible. The medical records may be subpoenaed for court, and it is usually not necessary to have the recording physician appear if that record is both legible and clear. A well-detailed record also assists in recalling the incident in the event that physicians or others are called upon to testify in court. One should avoid statements such as "virginal introitus" or "does not admit two fingers," which are terms which will easily confuse a courtroom and will almost certainly insure a court appearance.

11. FOLLOW-UP

Medical follow-up should include referral to the appropriate private physician or speciality clinic for follow-up of medical conditions. Follow-up should be arranged for surveillance of venereal disease, pregnancy, other gynecological sequelae, and emotional counseling.

12. PSYCHOLOGICAL ASPECTS OF RAPE

The psychological effects seem to depend on a number of factors.[17–21] Those that seem most important are the prerape personality, the child's age at the time of the rape, the relationship to the offender, and the quality of the response of the child's mother. The child's relationship with the father also figures in determining her future attitude to men. If the child is less mature, she is less capable of the emotional and physical coping that is necessarily called upon after rape. A confidence rape, one in which the child is enticed into sexual contact, tends to destroy the child's value system. It is difficult for the child to know who to trust if the trusted ones cannot be trusted. If the assault is more violent, this is more disruptive. The child often receives mixed messages from the mother. The mother is often unwilling to do much about the continued sexual contact between her daughter and her husband since that sexual contact lets the mother out of one of her marital responsibilities. This kind of behavior on the mother's part is confusing for the child. Her reaction may vary from one of competitiveness with mother by filling in for mother's duties in order to maintain the family to one of deep resentment of mothers' not being willing to risk to protect the child.

Sexual assault by a stranger, if handled supportively with understanding by the surrounding adults can lead to fewer long-term psychological effects since the basic supports are intact. If the situation, however, is mismanaged with inappropriate emotional reaction, then the child can be victimized and traumatized unnecessarily. Brother–sister incest has the highest incidence in the middle and upper socioeconomic levels and has been shown to be the least damaging long term to adult sexual adjustment. However, father–daughter sexual exploitation is more common in lower socioeconomic areas and produces a higher incidence of psychosexual disorder. Outside the topic of this book are mother–son and father–son sexual exploitation, both of which lead to a higher incidence of morbidity later in life.

Parental reaction is often related to the parents own sexual conflicts. Occasionally, parents can project their own sexual drives onto the child thus believing that the child should be punished because she in fact wanted the acts to occur. The other reason for confused parental responses is that many times the rapist is a family member or close friend, which leads to conflicted feelings of anger and guilt and may cause the parent not to act supportively toward the child.

As to the child's immediate reaction after the episode, a rape-trauma syndrome[20] has been described. The immediate reaction postassault is one of disruption of adaptation such that the patient regresses to a feeling of helplessness and child-like dependency with an increased suggestability to both positive and negative input. This reaction occurs in three phases. First, an acute reaction in the beginning with anxiety, agitation, and shock being the predominant emotions in addition to concern for practical issues. Second, a pseudoadjustment period where the patient returns to the usual pattern of function but continues with fears and nightmares and uses denial and depression as a mechanism of defence. Third, depression occurs later on as the defences breakdown requiring further integration of the experience after at least a month. Long-term sequelae can result if this syndrome is not cured. This long-term morbidity may lead to poor adult sexual relationships, ranging from avoidance of male contact and open homosexuality to hysterical seizures.

REFERENCES

1. Milling RN, Johnson MR: Changing attitudes and procedures in the crime of rape. *J So Car Med Assoc* 74:321–328, 1978
2. Scherzer LN, Lala P: Sexual offenses committed against children. *Clin Pediatr* 19:679–685, 1980
3. Hicks DJ: Rape: Sexual assault. *Am J Obstet Gynecol* 137:931–935, 1980
4. Soules MR, Steward SK, Brown KM, et al: The spectrum of alleged rape. *J Reprod Med* 20:33–39, 1978
5. Hayman CR, Lanza C, Fuentes R, et al: Rape in the District of Columbia. *Am J Obstet Gynecol* 113:91–91, 1972

6. Tilelli JA, Turek D, Jaffe AC: Sexual abuse of children. *N Engl J Med* 302:319–323, 1980

7. James J, Womack WM, Strauss F: Physician reporting of sexual abuse of children. *J Am Med Assoc* 240:1145–1146, 1978

8. Lipton GI, Roth EI: Rape: A complex management problem in the pediatric emergency room. *Pediatrics* 75:859–866, 1969

9. Robinson HA, Sherrod DB, Malcarney CN: Review of child molestation and alleged rape cases. *Am J Obstet Gynecol* 110(3):405–406, 1971

10. Dahlke MB, Cooke C, Cunnane M, et al: Identification of semen in 500 patients seen because of rape. *Am J Clin Pathol* 68:740–746, 1977

11. Silverman EM, Silverman AG: Persistence of spermatoza in the lower genital tracts of women. *J Am Med Assoc* 240:1875–1877, 1978

12. Soules MR, Pollard AA, Brown KM, et al: The forensic laboratory evaluation of evidence in alleged rape. *Am J Obstet Gynecol* 130:142–146, 1978

13. Bibbo M, Haenszel WM, Wied GL, et al: A twenty-five-year follow-up study of women exposed to diethylstilbesterol during pregnancy. *N Engl J Med* 298:763–767, 1978

14. Blye RP: The use of estrogens as postcoital contraceptive agents. *Am J Obstet Gynecol* 116:1044–1049, 1973

15. Anonymous: Postcoital contraception. *Br Med J*, October 23, 1976

16. Richmond JB: *Physician Advisory: Health Effect of the Pregnancy Use of Diethylstilbestrol.* Washington, DC, Department of Health, Education, and Welfare, October 4, 1978

17. Peters JJ: *The Psychological Effects of Childhood Rape.* Honolulu, Hawaii, American Psychiatric Association, May 9, 1973

18. Weeks RB: The sexually exploited child. *So Med J* 69:848–850, 1976

19. Peters JJ: Children who are victims of sexual assault and the psychology offenders. *Am J Psychother* 30:398–421, 1976

20. Schuker E: Psychodynamics and treatment of sexual assault victims. *J Am Acad Psychoanal* 7:553–573, 1979

21. Gross M: Incestuous rape: A cause for hysterical seizures in four adolescent girls. *Am J Orthopsychiat* 49:704–708, 1979

CHAPTER 9

Gynecologic Tumors of Adolescence

James D. Garnet

1. INTRODUCTION

Adolescence in the human female usually is defined as that era of life that begins at the onset of puberty and ends when adulthood is reached. A recent study[1] indicates that physical evidences of incipient puberty most commonly occur between 8.0 and 14.9 years and completion of such development is usually noted between 12.4 years and 16.8 years. Nubility may be defined as the status attained when sexual development is completed and the female is capable of procreation. Adulthood is somewhat more difficult to specifically define but usually may be considered to have been attained when nubility and a significant degree of psychologic maturity have been developed.

Gynecologic tumors of adolescence can be any neoplasm of the genital tract benign or malignant occurring during the era just defined. Benign neoplasms in themselves do not destroy the host; however, if malignant tumors remain untreated, destruction of the host is a foregone conclusion. Since the adolescent shows a propensity to develop any and all tumors encountered in adult life, albeit most are rare, one must at all times remain cognizant of these possibilities.

2. INCIDENCE

Unless specializing in pediatric gynecology, even the busy gynecologist sees relatively few genital tract tumors in this age group. These

James D. Garnet • Department of Obstetrics and Gynecology, Pennsylvania Hospital, Philadelphia, Pennsylvania 19107.

tumors are frequently rapidly growing and sometimes highly malignant so that there is no place for procrastination when such tumors are encountered. If the gynecologist has not had specialized training in pediatric oncology, immediate referral to such a specialist should be made.

The incidence of true neoplasia in children under 15 is reported by the American Cancer Society[1a] in their *1972 Cancer Facts and Figures* to be in the range of 11/100,000 or approximately 4000 each year. Only 2% of these are gynecologic malignancies with two-thirds arising within the ovary.

3. ETIOLOGY

When considering etiologic factors in certain of these tumors, it becomes apparent that almost any tissue in the body can be represented in the pluripotent cells of the female genital tract. Such cells thereby become the origin of rather bizarre and unusual neoplasms developing in this area of the female body. It has been postulated[2] that such foreign tissues are derived from misplaced blastomeres, which during early embryonic development have escaped the normally controlled organization of undifferentiated embryonic tissues. It has been demonstrated that these tumors may have a unicentric or multicentric origin.

An appreciation of how certain of the neoplastic growths arising in the genital tract of the younger adolescent can differ from those in the adult frequently will aid in the better management of these problems. Cancers seem to grow more rapidly, and because of limited space for tumor expansion in the pelvis, they will quickly become abdominal masses. Often a high degree of malignancy is apparent, possibly due to an inefficient immunologic mechanism for surveillance and early control of growth.[3] Therapeutic management requires recognition of the long-term effects that radical surgery and radical irradiation may have on the adolescent's further development.

4. DIAGNOSIS

The prevention of disease or its early recognition are concepts that in recent years have been encouraged and accepted by both physicians and patients. Thus, the yearly physical examination has become an important and integral part of gynecologic health care. This is particularly true in adult gynecology but considerably less so in the adolescent group of women. Because of this, there is a greater likelihood for the adolescent to consult with the gynecologist only after a significantly advanced pathologic abnormality has developed.

Of importance to making an accurate diagnosis is the need for a detailed history obtained from the patient. Only in the younger age group should the history be obtained from the mother. In most instances this information should be developed and obtained in an interview separate from the mother. Such an independent interview will significantly aid in developing a strong personal relationship between the adolescent and her physician. Should the patient sense that her gynecologist is simply relaying information to her mother, all trust and confidence can be lost.[4]

The physical examination should be complete, not just a pelvic examination. Thoroughness will assure the patient that her total health care is the physician's responsibility. This feeling strengthens confidence between the adolescent and her doctor.

Breast examination is a reassuring preliminary to the pelvic examination, again emphasizing thoroughness on the gynecologist's part. Just prior to the pelvic examination, learn whether the patient has had previous examinations or has used tampons. Such questions provide information concerning penetration of the vagina as well as relieving the patient's worries regarding the entire pelvic examination. Inspection of the vulva and vagina and the actual vaginal examination must be thorough as well as gentle. Since the speculum examination is usually the most uncomfortable procedure, it is well that it be preceded by the digital examination. Throughout the examination it is helpful, when the opportunity presents itself, to stress the normalcy of the structures thereby inhibiting tension and fostering relaxation. Negative comments, such as mention of a small vagina or uterus, should be replaced by positive assertions and demonstrations of normalcy. Abnormalities that have been noted can best be discussed with the patient after the examination has been completed and the patient clothed and in the more comfortable and relaxing environs of the consulting room. At this time in the life of the adolescent, the sensitivities demonstrated by the gynecologist can be important in helping her develop a mature self-image and feminine identity.

Appropriate laboratory studies, including routine cytologic studies, are fundamental to proper management. Additional studies and specific adjuvant examinations will be detailed as the individual tumors are discussed.

Before discussing the specific genital tumors common to adolescents, it is of utmost importance to recall the most common pelvic tumor encountered in the adolescent is the pregnant uterus. Pregnancy, either intrauterine or ectopic, must always be considered before definitive therapy for the pelvic mass is undertaken. The availability of an extremely sensitive pregnancy test in the form of the human chorionic gonadotropin β subunit blood test has greatly simplified this differential diagnosis during the very early stages of pregnancy. If this test result is positive, further differentiation between intrauterine and ectopic pregnancy is greatly facilitated by further ancillary studies, such as pelvic ultrasound examination, culdocentesis, and diagnostic laparoscopy.

Additionally, it must be recognized that clinical evaluation of a tumor cannot be relied upon alone to evaluate the neoplastic potential of many tumor masses. Biopsy and microscopic study are indicated in most instances.

5. VULVA

Proper evaluation of growths in the vulvar structures requires a basic appreciation of dermatologic diseases, since most of the lesions affecting the integument in general can be seen on occasion in the skin of the vulva. Of the nonneoplastic disorders common to the vulva are the cysts of the canal of Nuck, paraurethral cysts, and Bartholin gland cysts. The first two of these are treated by local excision and the third, in most instances, by marsupialization.

Benign neoplastic structures include the hydradenoma,[5] a cystic tumor of the sweat glands, which is treated by local excision. Those tumors arising from mesodermal tissue are the fibroma, lipoma, leiomyoma, lymphangioma, and hemangioma. Those that are solid tumors are treated by local excision, while the lymphangioma and hemangioma can be followed by careful expectant observation. Should bleeding, ulceration, or distortion due to attainment of a large tumor mass occur, excision becomes indicated. Recently these vascular masses have been destroyed utilizing the argon laser.[6] This technique has gained acceptance since bleeding is readily controlled and minimal scarring results. An extremely rare tumor derived from mesodermal tissue is the granular cell myoblastoma,[7] a slowly growing, poorly encapsulated, firm nodule, usually benign, which can be eliminated by wide local excision.

Condyloma accuminata has become the most common epithelial lesion among adolescent girls to such a degree that this virus-induced growth has reached epidemic proportions in many areas. It not only involves the vulva but may also be found in the vagina, on the cervix, at the urethral meatus, and in the anal canal. Careful local hygiene utilizing drying powders will sometimes cause regression. Application of 25% podaphyllin can be tried on small lesions; however, it is seldom successful or safe on larger lesions. It should not be used during pregnancy, since it is absorbed into the general circulation with serious and sometimes fatal effects on the fetus. Surgical excision by electrocoagulation or cold knife is used for larger lesions. Such large growths respond extremely well to carbon dioxide laser[8] vaporization, which has the advantages of readily controlled bleeding and with minimal scarring and postoperative pain.

Precancerous epithelial changes of the vulvar skin are uncommon in the adolescent but do occur. Skin changes associated with dysplasia or carcinoma *in situ* are variable and may appear as white areas, brownish pigmented areas, or atrophic thinned and shiny areas of skin. These dystrophic lesions may be hypertrophic, atrophic, or mixed, and any of these three forms may be dysplastic in nature. Biopsy of abnormal areas of skin is indicated. Malignant precursors, if found in the adolescent warrant a conservative approach consisting of local excision of the lesion by cold knife or laser[8] techniques. Careful long-term follow-up is imperative.

Malignant tumors of the vulva are extremely rare in the adolescent, but if they are encountered, radical vulvectomy with bilateral inguinal and femoral lymph node dissection is the procedure of choice. If the superficial nodes are positive for metastatic tumor, deep node dissection becomes indicated.

Basal cell carcinoma is amenable to wide local excision.

6. VAGINA

An imperforate hymen will occasionally be observed at the time of inspection of the vulva. When associated with a soft bulging of the hymen prior to the onset of menstruation, a collection of watery mucus may be present. If a cystic bulge occurs after menstruation has been established, hematocolpos will be found. Simple incision of the hymen will correct both of these conditions. Other nonneoplastic lesions are the congenital Gartner's duct cysts and epithelial inclusion cysts following vaginal delivery in the older adolescent.

Vaginal adenosis, a glandular epithelium similar to the endocervical epithelium, may be congenital or acquired. Congenital adenosis, arising from remnants of the mullerian system which during embryonal life have been trapped beneath the epithelial surfaces, is extremely rare. In 1970 Herbst and Scully[9] described an acquired form of adenosis involving adolescent offspring of mothers who ingested diethylstilbestrol (DES) during pregnancy. Since then, an increasing incidence of these lesions has been observed involving both the vagina and cervix. Occurrences of clear cell and squamous cell carcinoma have been infrequently reported. Thus far, few cases of clear cell carcinoma have been prospectively identified in patients with previously demonstrated benign adenosis. The predisposition of vaginal adenosis to undergo malignant change requires further clarification. All girls or women with a history of intrauterine exposure to DES should have careful physical examination including cervical and vaginal cytologic study and colposcopy starting at menarche or by age 14 years. Subsequent follow-up at 6-month intervals with repeat colposcopy at 1- to 3-year intervals is indicated. Should there be unusual bleeding, staining of blood, or vaginal discharge, examination is immediately indicated without regard to age of the patient.

Of the few primary carcinomas of the vagina occurring during adolescence, the great majority are adenocarcinomas derived from the mesonephric remnants. With the lesion localized within the pelvis, radical vaginectomy and hysterectomy with pelvic lymphadenectomy is indicated. Ovarian conservation is imperative.

Another of the malignant tumors arising in the vagina during childhood and early adolescence are the sarcomatous growths of which sarcoma botryoides[10,11] is the most common. This is best described as an embryonal myosarcoma arising from the subepithelial tissues of the vagina. The tumors are multicentric, involving the lower vagina and extending upward to the cervix and downward to the vulva. The tumor often has a grape-like appearance with rapid growth and expansion in the subepithelial tissues. Most commonly the cells are densely packed and very poorly differentiated. Occasionally this tumor will occur as a well-differentiated rhabdomyosarcoma. With localization of the tumor, complete vaginectomy and hysterectomy is indicated. If extension beyond the confines of the subepithelial tissue occurs, exenterative procedures are required. More recently adjunctive chemotherapy has shown value in controlling these tumors. Combinations of actinomysin, Adriamycin®, vincristine, and Cytoxan® have been effective.

7. UTERUS

7.1. Cervix

Most of the benign tumors involving the cervix uteri in the adult may also be encountered in the adolescent. Included in this group are endocervical polyps, condyloma acuminata, endometriosis, acquired adenosis, hemangiomas, and squamous papillomas. Polyps are readily recognizable since most are pedunculated and arise from the endocervical mucous glands. Treatment is usually avulsion of the polyp and electrocauterization of the pedicle base. Condylomas are recognized and treated as described when involving the vulva or vagina. Endometrial implants on the exocervix are brilliant red, slightly elevated lesions seldom more than 3–5 mm in diameter. They tend to exhibit intermittent spontaneous bleeding or associated with coitus or other irritations to the cervix. Biopsy diagnosis is indicated and treatment by electrocauterization or cryocauterization suffices. DES-acquired adenosis is frequently identified by the cockscomb anterior deformity of the cervix or the collar of adenosis involving the entire periphery of the cervix. Treatment of these lesions is seldom indicated; however, should there be a bothersome mucous discharge, destruction of the extensive glandular elements can be accomplished by cryocauterization or laser beam therapy. Hemangiomas and squamous papillomas are managed as when occurring on the vulvar skin.

Numerous studies have established the fact that sexual activity is a major factor in the genesis of cervical carcinoma. There appears to be an association of cervical cancer with factors related to an early age of first intercourse. This hypothesis[12] proposes that during adolescence the cervical epithelial cells are especially vulnerable to carcinogens. The implicated carcinogens include chronic cervicitis, repeated cervical infection, hormonal imbalance, smegma, trauma, or sperm DNA. The association of cervical cancer with early coitus is another observation. This hypothesis stresses the multiplicity of sexual partners, not only of the woman herself, but also of her husband. It is proposed that the malignant precursors, which may ultimately progress to frank malignancy, are induced by a sexually transmitted infection. The herpes virus Type II, which causes genital herpes, has been implicated as the specific infectious agent. There is considerable evidence that cervix cancer may be venereally transmitted. A study[12] in which the incidence of cancer of the cervix by year of birth is coupled with the incidence of gonorrhea by year of birth demonstrates a close association between these two diseases. Such evidence makes it difficult to escape the fact that cancer of the cervix is a venereal disease.

Despite the sexual freedom encountered in the adolescent over the past decade, there has been a steady decline in the incidence of invasive cancer of the cervix. A significant factor could well be the more widespread use of the Papanicolaou smear with early recognition and treatment of the benign precursor lesions. With the increasing frequency that precursor lesions are being encountered in teenage girls, it is urgent that all those who are having sexual intercourse should be followed by annual cytologic studies. Any cytologic atypias or dysplasias including carcinoma *in situ* (CIS) demonstrated on the smear require further evaluation by tissue biopsy and histologic diagnosis. With the resurgence of interest in colposcopy[13] in the United States, accurate evaluation of the abnormal cytologic smear has been enhanced. When this examination is performed by a physician thoroughly experienced in the technique, a large proportion of cone biopsies of the cervix can be eliminated thereby avoiding many of the serious complications associated with this procedure. An accurate tissue diagnosis can be obtained by examination of small biopsies taken from abnormal areas on the cervix as identified by colposcopic examination. Those lesions not amenable to colposcopic identification must be evaluated by the histologic study of the cone biopsy. These precursor lesions are classified in four groups; mild, moderate, and severe dysplasia and CIS. The mild and moderate forms are usually treated by cauterization, preferably cryocauterization or laser beam[14,15] destruction. Severe dysplasia and CIS are treated by therapeutic conization of the cervix if the patient desires to preserve future childbearing capabilities. If there is no desire for future pregnancies, total hysterectomy with preservation of the ovaries is indicated. All patients must continue to be followed with cytologic studies at 3- to 6-month intervals.

Although uncommon, both squamous cell and adenocarcinoma of the cervix are known to occur in the adolescent. In the past, adenocarcinomas of mesonephric origin were more common; however, in recent years there is an increasing number of squamous cell tumors being encountered in adolescents. Treatment of these invasive lesions is largely dependent upon the stage of the disease at the time of diagnosis. The earliest are those with minimal microscopic invasion extending just below the basement membrane. Such microinvasive tumors can be safely treated by simple extrafascial hysterectomy. Tumors extending deeper than the microinvasive lesions but limited to the cervix or upper vagina can best be treated by radical hysterectomy and pelvic lymphadenectomy. The ovaries should be conserved in these young women. With extension of the tumor beyond the cervix, radical radiation therapy becomes indicated. A very small number of lesions that spread to the bowel and/or bladder with no lateral spread are best treated by exenterative procedures if feasible.

Sarcoma botryoides may also arise from the cervical connective tissue. The principles of management are similar to those described when the tumor arises from the vagina.

7.2. Corpus

Benign tumors arising from the uterine corpus are leiomyomas, polyps, and adenomyosis, all of which are quite rare. Diagnosis is essentially as in the young adult, and therapy is such as to preserve ovarian function and fertility whenever possible.

Fortunately the invasive forms of endometrial cancer are rare in the adolescent female; however, the so-called premalignant or precursor changes in the endometrium are encountered more commonly. Such hyperplastic changes[16] result from continuous unopposed estrogen stimulation of the endometrial glands. The absence of progestational maturation of the glandular elements allows progressive estrogen hyperplasia to proceed to a degree where invasive malignancy threatens. Endocrine disturbances that cause a persistent or prolonged episode of anovulation is the most common cause for the continuous estrogen stimulation of the endometrium. Other less common causes are the polycystic ovary syndrome and estrogen-secreting ovarian tumors. Irregularities of bleeding that appear to be dysfunctional in character is the most common symptom. Diagnostic curettage of the endometrium will reveal the diagnosis and also serves to evaluate the therapeutic effects. Certain endocrine clues, such as anovulation, infertility, obesity, and hirsutism, suggest the possibility of polycystic ovaries. Ultrasonography and laparoscopy will clinch the diagnosis of polycystic ovaries or demonstrate the presence of an ovarian tumor suspect for estrogen secretion. Cyclic progestational therapy will prevent progression of the hyperplastic changes and when withdrawn will cause sloughing of the endometrium. Medication for the stimulation of ovulation may be indicated at times, and standard endocrine or surgical treatment utilized when indicated, for correction of the polycystic ovary syndrome. Resection of an estrogen active ovarian tumor is mandatory.

As in the adult, trophoblastic disorders[17] may be associated with adolescent pregnancies. Both the benign forms and the malignant choriocarcinoma are readily diagnosed and therapeutic results followed by the use of the human chorionic gonadotrophin β subunit blood test. This endocrine study is an accurate tumor marker for this disease. Treatment and follow-up of this disease can often be quite complicated. Salvage rates can be extremely high if the disease is properly treated and is often best managed when handled at a center where these lesions are seen with some frequency.

The diagnosis and treatment of the fortunately infrequent endometrial carcinoma and uterine sarcoma is as detailed for the tumors when occurring in the adult.

8. FALLOPIAN TUBE

The most commonly encountered benign tumors of the fallopian tubes are the parovarian cysts arising from the para-oophoron. These may attain a large size and require resection.

Malignancies of the tubes are so rare as to be almost nonexistant.

9. OVARY

As in the adult, ovarian tumors in the adolescent represent a wide spectrum of disease, ranging from benign functional cystic ovarian enlargement to highly malignant solid tumors. The ovary is the most common site for gynecologic neoplasms occurring in adolescence. The incidence usually reported is 1–2% of all tumors during the adolescent years. These tumors may occur at any time in childhood or adolescence but tend to be more frequent between ages 10 and 14 years.[3] This finding suggests a relationship of the endocrine changes during menarche and an increased susceptibility of ovarian tissues for neoplastic changes.

Classification of ovarian tumors[18] according to their histogenesis greatly simplifies and clarifies what otherwise can become a group of confusing and greatly varying tumors. The three tissue sources of these tumors include the mesothelium, from which arise the so-called germinal epithelial tumors and the mesoderm, which gives rise to the gonadal stromal tumors, i.e., granulosa–theca cell, Sertoli–Leydig cell, hilus cell, adrenal cell, and gynandroblastoma. Also arising from the mesoderm are the ordinary connective tissue tumors, i.e., fibroma, angioma, myoma, and sarcoma. The third group arises from the germ cell and includes the dysgerminoma and gonadoblastoma, the teratomas either mature or immature (embryonal), and the extraembryonal yolk sac tumors (endodermal sinus tumors).

Unfortunately there are no early manifestations of this disease whether benign or malignant. The usual findings of abdominal enlargement, pain, and a palpable mass indicate advanced disease with spread beyond the ovary already having occurred. Means of early detection are extremely limited so that early diagnosis becomes primarily a matter of chance. In the younger adolescent, ovarian tumors are associated with problems less frequently seen in the older adolescent or the adult. As growth of the tumor occurs, it rapidly becomes an abdominal mass due to limited pelvic space for expansion. This creates greater pelvic and intraabdominal pressure with concurrent pain and dyspnea. Symptoms and physical findings are directly related to the rapidity of tumor growth, location, degree of malignancy, and presence or absence of endocrine activity. Complications to be recognized include torsion, rupture, hemorrhage, and infection.

In addition to complete medical and gynecologic history, a careful abdominal, pelvic, and rectal examination must be performed. The rectal examination may at times indicate that the pelvis is free of tumor; however, this finding does not rule out an ovarian tumor, which may frequently present as an abdominal mass in even the earlier stages of tumor growth. It is often almost impossible to palpate normal ovaries in children and younger adolescents. Therefore, it can usually be assumed that if an ovary is enlarged on palpation, it is probably abnormal. In the older adolescent who is menstruating regularly, a cystic enlargement of the ovary that is less than 6 cm in diameter may safely be watched for 6–8 weeks to determine whether the cyst fluid is absorbed and the cystic mass disappears. If this occurs, one can be certain that it was nothing other than a physiologic cyst[19] and nothing further need be done. If the mass remains or enlarges, further studies become indicated.

Baseline blood counts, urinalysis, and blood chemistry should be obtained. A flat plate X-ray study of the abdomen may be helpful in determining the presence of characteristic calcium deposits that may be present in certain of these ovarian tumors. Ultrasound studies of the pelvic organs now play an important role in the diagnosis of ovarian tumors. In some instances where the diagnosis is still in question, laparoscopic visualization of the pelvic organs becomes indispensible. Ovarian hormone assays may in some instances be an aid in diagnosis as well as the α-fetoprotein titer, which may be elevated in the extraembryonal endodermal sinus tumors as well as in certain embryonal carcinomas.

The clinical picture and physical findings usually indicate the diagnosis. However, occasionally intravenous pyleograms and X-ray studies of the gastrointestinal tract are required to differentiate certain other pelvic diseases.

9.1. Germinal Epithelial Tumors

In the adult woman, the epithelial tumors comprise approximately 90% of ovarian tumors, while in the adolescent they account for only 28% of the ovarian tumors.[20] Included in this group of tumors are the serous, mucinous, endometroid, clear cell, and the mixed-epithelial tumors. Histologically, they may be identified as being completely benign, borderline or very-low-grade malignancy, or unequivocally malignant tumors. Should the tumor be benign, unilateral ovariectomy with conservation of the opposite ovary is indicated, if the wedge resection taken from that ovary shows no evidence of tumor involvement. If the tumor is malignant but completely encapsulated, unilateral with no evidence of metastatic disease, and cytologic study of the peritoneal fluid is negative for tumor cells, again unilateral ovariectomy with bisection and biopsy of the contralateral ovary is sufficient treatment. Nonetheless, these young patients must be followed extremely carefully throughout the remainder of their lives for evidence of tumor development in the conserved ovary. If there is any evidence of spread beyond the ovary, a total hysterectomy, bilateral salpingoophorectomy, appendectomy, and omentectomy should be performed. Peritoneal fluid or washings should be obtained for cytologic study. Adjuvant chemotherapy is probably indicated in all patients who have shown extension of tumor beyond the ovary.

9.2. Gonadal Stromal Tumors

Histogenetically this group of tumors is derived from the cells of the ovarian mesenchyme or mesoderm. These cells give origin to the stromal tissue of the gonad. Granulosa–theca cell tumors may give rise to a feminine expression and the Sertoli–Leydig cell tumors to a masculine expression. These tumors[21] are not at all common in the adolescent, but when they do occur they may infrequently produce either an estrogenizing or masculinizing effect.

The tumors of theca cell origin, referred to as thecomas, are almost always benign and treatment consists of resection of the tumor or unilateral ovariectomy.

The granulosa cell tumor is usually of low-grade malignancy and most often unilateral in nature. When the tumor is unilateral with no evidence of spread beyond the ovary, unilateral salpingoovariectomy is indicated. Where there is evidence of metastatic spread beyond the ovary the surgical procedure will usually include bilateral salpingoophorectomy and hysterectomy. Since this tumor is radiation sensitive, radiation therapy should be given to patients who have proved metastatic spread.

Among the male cell-type tumors, the mixed Sertoli–Leydig cell tumors, often referred to as the arrhenoblastomas, are the most significant and may occasionally be seen in the adolescent female. This tumor may or may not cause defeminization and then masculinization of the patient. Indications for conservative surgery[22] are the same as for the granulosa cell tumors.

Gynandroblastomas are infrequent gonadal stromal tumors and are composed of cells from both male and female cell types in about equal proportions. This rare tumor is managed in the same fashion as are the granulosa cell tumors.

Since these tumors frequently have delayed recurrence rates, which may extend well beyond 5 years, follow-up[23] should be of long-term duration.

The ordinary connective tissue tumors may arise from the ovarian mesoderm and include the fibromas, angiomas, myomas, and sarcomas. All are extremely rare except the fibroma, which is benign and may at times be associated with tumors arising from the epithelial cell origin and are referred to as fibroadenomas. Malignant variants of this tumor may infrequently be encountered.

9.3. Germ Cell Tumors

9.3.1. Teratomas

The germ cell tumors of the ovary are more commonly found in children and adolescents rather than adults. Of these tumors, the teratomas are by far the most commonly encountered subgroup. There are two major forms of ovarian teratomas: the mature or adult type and the immature or embryonal type. Of the mature teratomas, the benign cystic teratoma[24] accounts for more than one-third of ovarian tumors in the adolescent. Such tumors, commonly but incorrectly referred to as dermoid cysts, are composed of all three germinal layers although epithelial structures predominate. Nuclear sex chromatin bodies in ovarian teratomas are uniformally positive and therefore indicate the presence of the XX sex chromosome. This has been confirmed by chromosomal studies.[3]

On pelvic examination, the cystic teratoma is palpated as a rounded, nontender, smooth, heavy, mobile mass, which is often anterior to the uterine body. In children and in the younger adolescent, the mass is usually palpated above the true pelvis. X-ray studies of the pelvis and abdomen will demonstrate teeth or other calcified tissues and the halo sign in better than 50% of the cases. Ultrasonic study of the mass is frequently quite accurate in identifying this tumor.

Frequently the adolescent will seek medical attention only after the onset of pain due to twisting, hemorrhaging, perforation, or infection. Rarely, a localized leak of the cystic contents will cause a pelvic peritonitis, and the tumor may become densely adherent to the surrounding structures of the pelvis. A differential diagnosis must then be made between malignancy, tuberculosis, and endometriosis.

On microscopic examination, all types of mature ectoderm, mesoderm, and endodermal elements may be identified. The ectodermal elements, which are usually most common, are represented by skin, hair follicles, and sebaceous and sweat glands. Large quantities of sebaceous material frequently are present as well as teeth, bone, cartilage, fat, fibrous tissue, and muscle representing structures derived from the mesoderm. Structures of endodermal origin, such as respiratory tract tissue, thyroid tissue, and gastrointestinal mucosa, may be identified. When the teratoma is monodermal and composed primarily of thyroid tissue or if there is clinical evidence of thyrotoxicosis, the tumor is referred to as struma ovarii. Another uncommon variant of the cystic teratoma is the ovarian carcinoid. This tumor is often monodermal and may show the clinical features of the carcinoid syndrome, which are due to the formation of excessive quantities of serotonin, 5-hydroxytryptamine.

Malignant change occurring in the cystic teratoma is rare, involving only 1% of these tumors. Bilaterality of the benign cystic teratoma is in the range of 12%.

When the tumor lies free in the peritoneal cavity, careful enucleation and conservation of the remaining ovarian tissue can usually be accomplished. Reconstruction of the residual normal ovarian tissue should be performed whenever possible. Careful examination should be made of the contralateral ovary by palpation and usually with bisection and biopsy to determine the presence of small foci of disease. If present, the areas should be resected and the ovary reconstructed and preserved. Areas in these tumors that are highly suspect for malignant change should be submitted for frozen section study.

Since microscopic areas of teratomatous changes within the preserved ovarian tissue may go unrecognized, it is well to explain to the patient the possibility of recurrence of this benign condition. Careful follow-up of these patients is imperative.

Immature or embryonal teratomas are usually solid and composed of both embryonal and mature tissues, which are derived from all three germ layers. A commonly accepted theory of histogenesis is that of spontaneous growth of a primitive unfertilized ovum. The mixtures of embryonal and mature tissues within these tumors may vary greatly. The degree of malignancy of these tumors is primarily dependent on the quantities of embryonal tissue being present. Those tumors in which the cells are well differentiated and in which there are only rare small foci of embryonal tissues have the best prognosis. Those showing large quantities of embryonal tissue and extreme cellular atypicality are highly malignant and have an extremely poor prognosis. It has become apparent that the prognosis for these tumors is also dependent upon the predominant cell type present. An abundance of neural tissue carries a much better prognosis than does a tumor containing choriocarcinoma or endodermal sinus components.

Approximately 10% of these tumors are bilateral. When this tumor is unilateral, well encapsulated, and with no evidence of spread beyond the ovary, radical extirpation of the pelvic organs has shown no advantage over unilateral oophorectomy. However, when the tumor is bilateral or if there is spread beyond the ovary, total abdominal hysterectomy and bilateral salpingoooophorectomy, appendectomy, and omentectomy are indicated. In either situation, peritoneal washings should be taken for cytologic study. In recent years, intensive multiagent chemotherapy following definitive surgery has improved the salvage rate when used prophylactically, in the presence of persistent tumor, or when recurrences develop. Drugs that have been used for this purpose include actinomycin-D, 5-fluorouracil, cyclophosphamide, vincristine, and Adriamycin. Although chemotherapy administered to women of childbearing age may cause ovarian dysfunction and sterility, when discontinued there will often be a return of fertility.

9.3.2. Endodermal Sinus Tumors

The third type of teratoma encountered in the ovary is the extraembryonal form. The endodermal sinus tumor is a rare tumor but more common than the extraembryonal choriocarcinoma.

The endodermal sinus tumors[25] are thought to arise from the yolk sack endoderm and may exhibit a great number of blastocyst-like yolk sack vesicles and are referred to as polyembryonic embryoma. There also may be the polyvesicular vitelline tumor with its characteristic histologic pattern occasionally found as a predominant part of vitelline tumors in the ovary. These tumors commonly occur in the younger adolescent patient. In most instances this is a highly malignant tumor but when unilateral and well-encapsulated treatment by unilateral oophorectomy is as effective as total hysterectomy and bilateral salpingoophorectomy. Radiation therapy has little to offer since the tumor is radioresistant. The role of chemotherapy in the treatment of this tumor is becoming established. With multiagent chemotherapy[26] as described for treatment of immature teratomas, there has been a significant improvement in the salvage rate in these patients.

9.3.3. Choriocarcinoma

Primary choriocarcinoma arising in the ovary is an extremely rare tumor, which is highly malignant. This tumor may arise as a primary extraembryonic tumor as well as from a teratoma. When encountered in the ovary of the older adolescent, it is most commonly the result of metastases from a primary uterine choriocarcinoma. The tumor in the ovary is grossly the same as that seen elsewhere and is characterized by extensive hemorrhage, ulceration, and necrosis. In this form of choriocarcinoma, as in the primary uterine choriocarcinoma, the human chorionic gonadotrophin β subunit is utilized as a tumor marker for diagnostic purposes as well as for evaluation for therapy. Indications for operative intervention include resection of the primary tumor in the ovary as well as areas of metastatic tumor that do not respond to chemotherapy. However, multiagent chemotherapy is required to eliminate most foci of metastatic disease.

9.3.4. Dysgerminoma

Of the malignant germ cell tumors, dysgerminoma[27] is the most common with 40–45% of all these tumors occurring in patients under 20 years of age. The tumor is an embryonal type resembling the sexually undifferentiated germ cells of the early gonad. It has been reported that the origin of this tumor is probably linked to the continuing proliferation of unencapsulated germ cells and the associated stimulation of the surrounding stromal or ovarian mesenchymal cells. The majority of these tumors have a homogenous pattern, but occasionally a dysgerminoma can coexist with various structural combinations of other germ cell lesions. The male counterpart of this tumor is designated as a seminoma.

The tumor is usually smooth or lobulated and surrounded by a dense capsule. It has a rubbery consistency and may vary in size from a few centimeters to a huge tumor filling the pelvis and the lower abdomen. Histologically, the pure dysgerminoma consists of rather large round cells with dark-staining nuclei and stroma dividing the tumor into nests of cells. The stromal septi are heavily infiltrated with lymphocytes and giant cells. The tumor involves both ovaries in approximately 10% of patients. Certain common characteristics are noted in the patients with ovarian dysgerminoma. Among these are the youth of the patient and the lack of any specific signs or symptoms other than those commonly associated with an expanded intrapelvic tumor. The tumors are generally biologically inert unless an incidental choriocarcinoma coexists with the dysgerminoma. The age of the menarche falls within the normal range and menstrual irregularities are only rarely encountered. A small percentage of the patients will have coexisting developmental anomalies of the genital structures. However, it must be emphasized that the majority of dysgerminomas occur in apparently normal women.

Much disagreement exists as to what may be considered the best mode of therapy for the patient with dysgerminoma. In view of the youth of most of these patients, preservation of reproductive capabilities constitutes one of the major causes for these widely divergent views. Additionally, controversies concerning the proper form of treatment are related to incorrect diagnosis. The salvage rate with a pure dysgerminoma localized to one ovary ranges between 80 and 90%. However, if rupture of the tumor has occurred, there is a 50% recurrence rate in the contralateral ovary if it is retained. With a unilateral and well-encapsulated tumor, the contralateral ovary and the uterus can be conserved for childbearing purposes if at the time of exploration a biopsy of the remaining ovary shows no tumor to be present, if the paraaortic lymph nodes are negative for metastatic tumor, and if pelvic cytology is negative for tumor cells.

When the tumor has extended beyond the ovaries, or if both ovaries are involved by tumor, total abdominal hysterectomy and bilateral salpingoophorectomy with appendectomy and omentectomy becomes indicated. Additional radiation therapy is then given to the total pelvis and paraaortic lymph node areas. In the younger patient where total abdominal radiation is indicated, one must accept the fact that the patient may not have achieved full growth and that harm may be done to structures, such as the bone. The tumor tends to recur early and the greatest incidence is within the first 3 years following initial therapy. The patient should be seen for follow-up examination every 3–4 months during this time, and should there be any suspicion of tumor recurrence, exploratory laparotomy and biopsy is recommended so that early adjunctive therapy can be immediately instituted. This tumor is highly radiosensitive and early recurrences frequently can be irradiated with eradication of the tumor. Adherence to these therapeutic principles will permit preservation of reproductive capability without added risk to the patient.

9.4. Gonadoblastoma

The gonadoblastoma is a rare tumor composed of both germ cells and gonadal stromal cells. Most patients with gonadoblastoma are intersexual with a phenotypic female habitus, amenorrheic, and may be virilized. This tumor frequently arises in the dysgenetic gonad. The patients often show signs of Turner syndrome and are sterile. In most cases that have been studied, the nuclear pattern[28] has been negative, 46XY, or they may have a sex chromosome mosaicism XO–XY.

These tumors are bilateral in approximately one-third of the cases. A characteristic feature of the tumor is the frequency of speckled areas of calcification, which may readily be demonstrated on X-ray examination. Microscopically,[28] there are masses of large germ cells with intermingled granulosa-like cells. The tumor may also have cells present that are suggestive of Leydig cells as well as those reminiscent of dysgerminoma cells or of endodermal sinus tumor cells. The pure gonadoblastoma is not a highly malignant tumor, but the malignant potential will depend upon the presence or absence of a more malignant germ cell variant. Since this tumor is associated with the dysgenetic streak gonads, such contralateral streaks should always be removed because of their potential for developing a malignancy. These patients may then safely receive estrogen–progesterone substitutional therapy.

An atlas of the lesions described follows this discussion (Figs. 1–14).

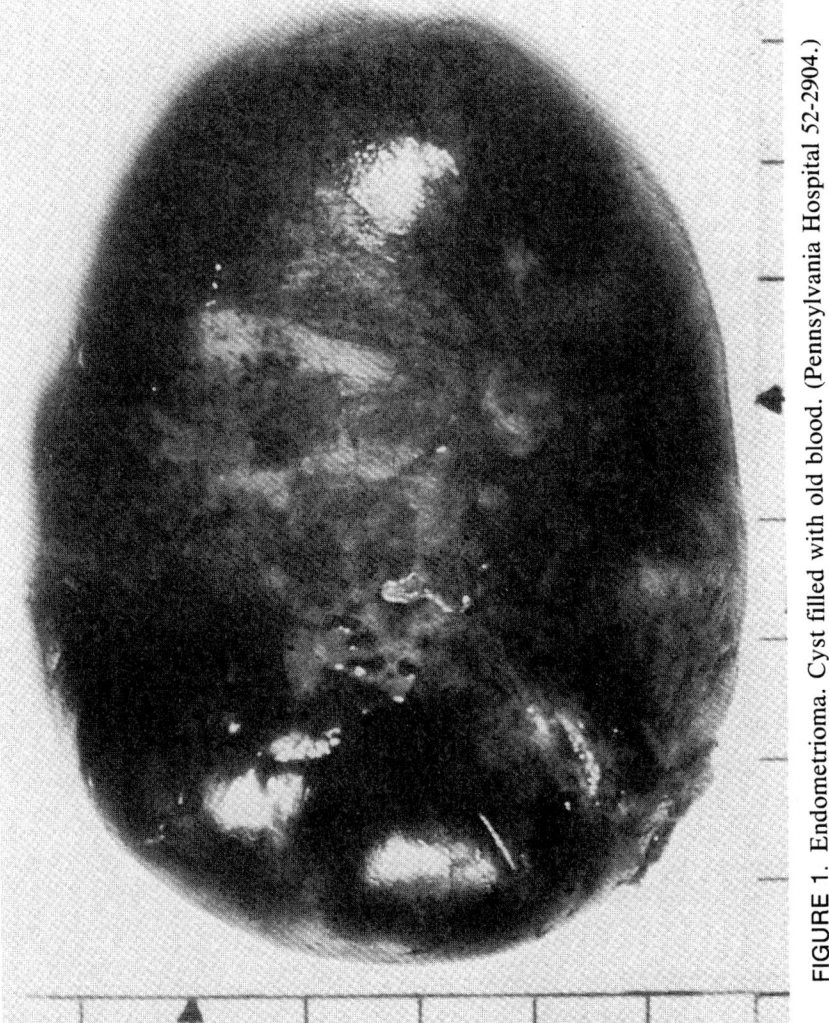

FIGURE 1. Endometrioma. Cyst filled with old blood. (Pennsylvania Hospital 52-2904.)

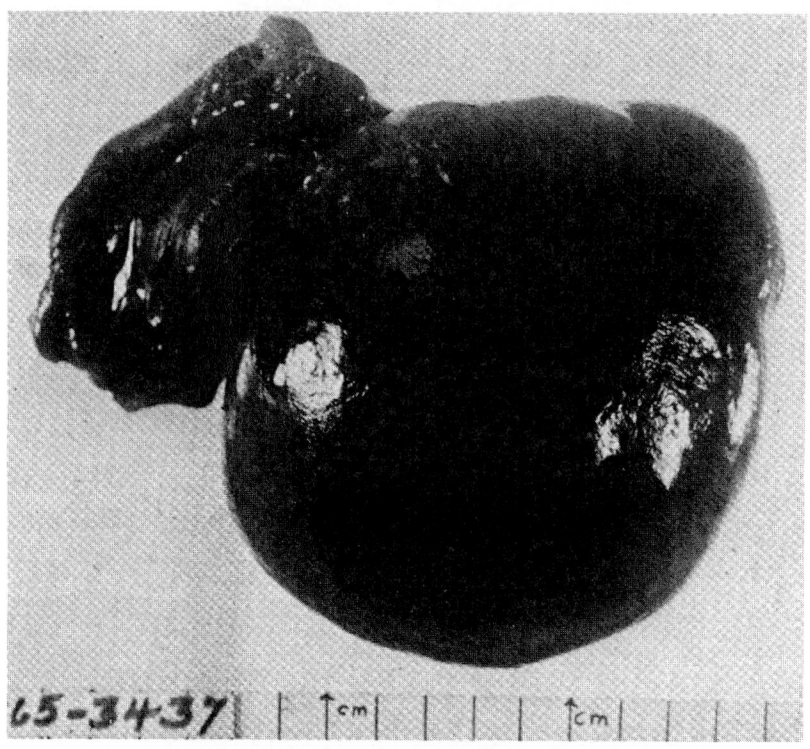

FIGURE 2. Fallopian tube and ovary with serous cystadenoma—torsion and infarction. (Pennsylvania Hospital 65-3437.)

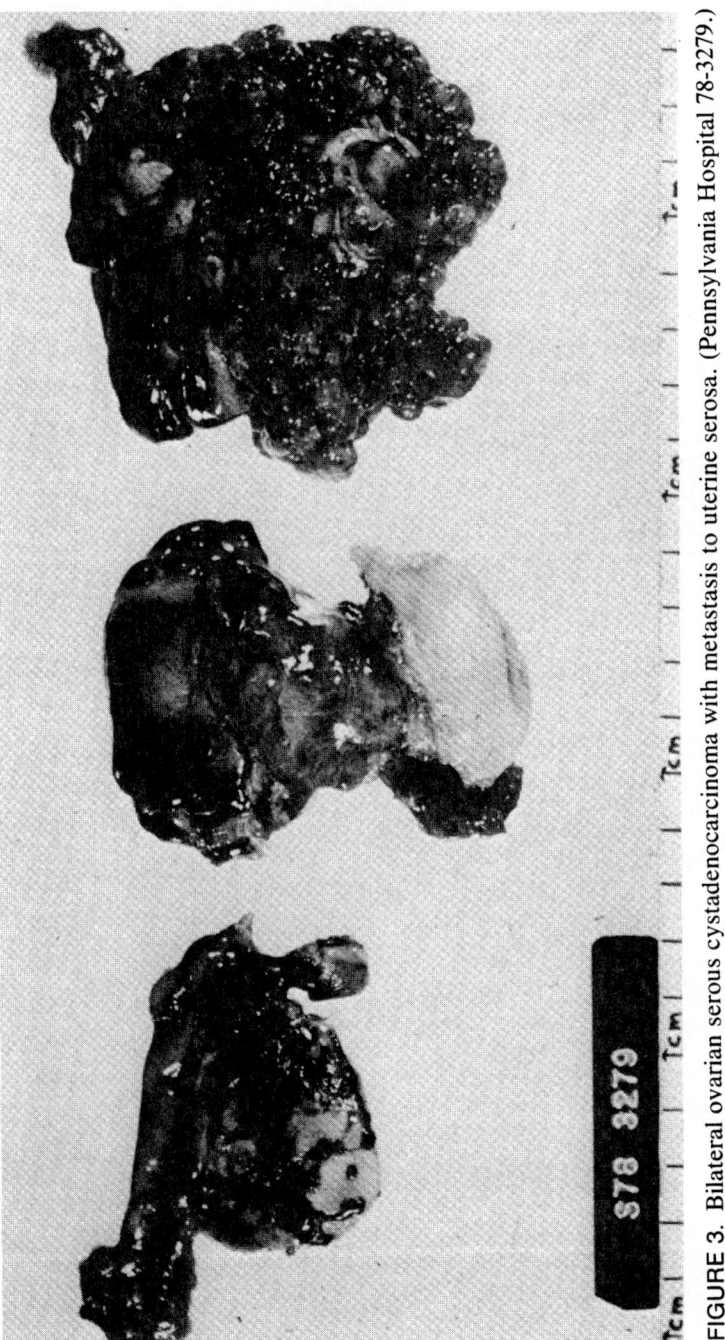

FIGURE 3. Bilateral ovarian serous cystadenocarcinoma with metastasis to uterine serosa. (Pennsylvania Hospital 78-3279.)

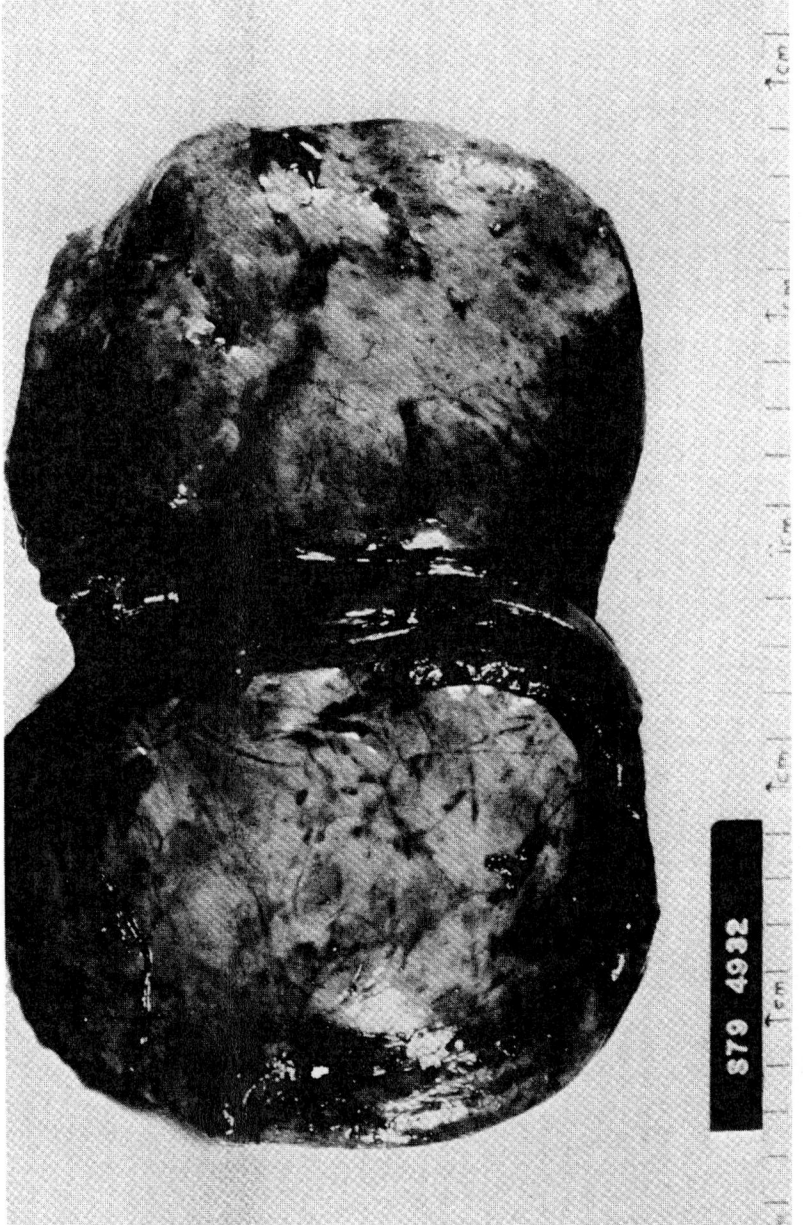

FIGURE 4. Granulosa cell tumor. Smooth external surface. (Pennsylvania Hospital 79-4932.)

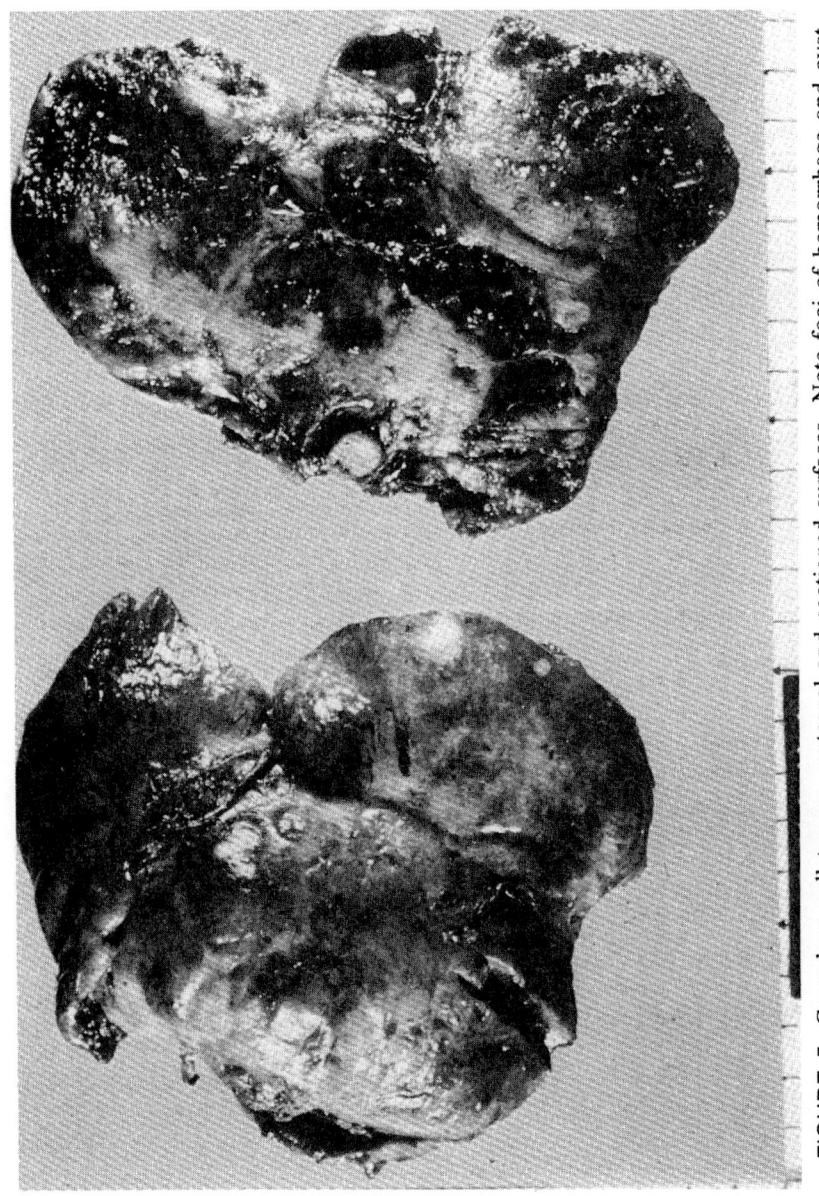

FIGURE 5. Granulosa cell tumor—external and sectioned surfaces. Note foci of hemorrhage and cyst formation. (Pennsylvania Hospital 54-2971.)

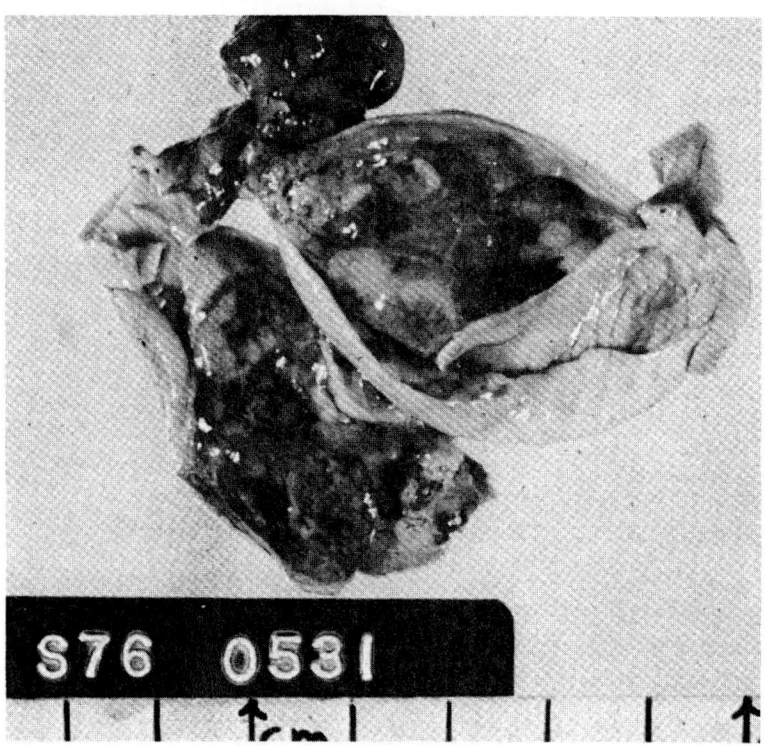

FIGURE 6. Sertoli–Leydig cell tumor. Predominantly solid tumor with single large cystic space. (Pennsylvania Hospital 76-531.)

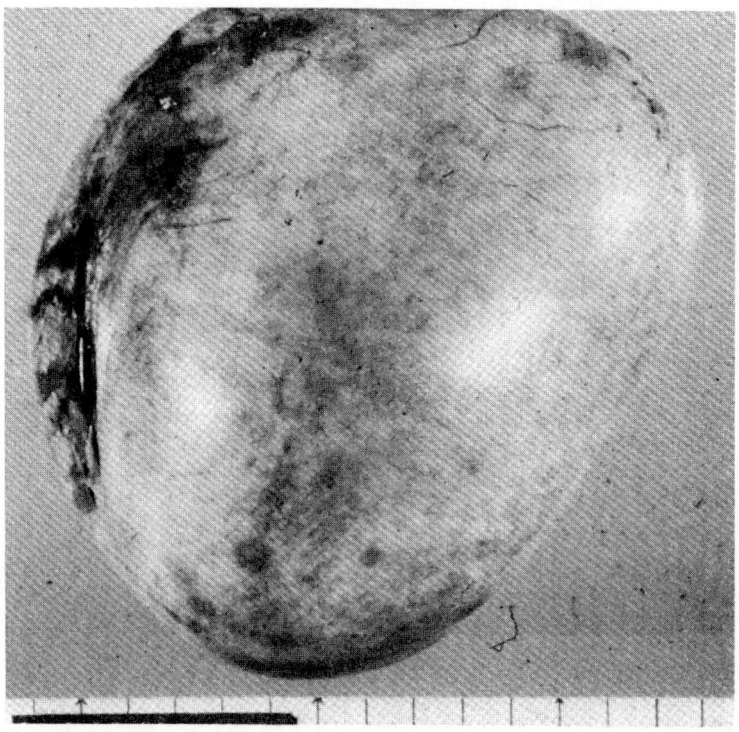

FIGURE 7. Benign cystic teratoma. Characteristic smooth, gray-white external surface. (Pennsylvania Hospital 57-534.)

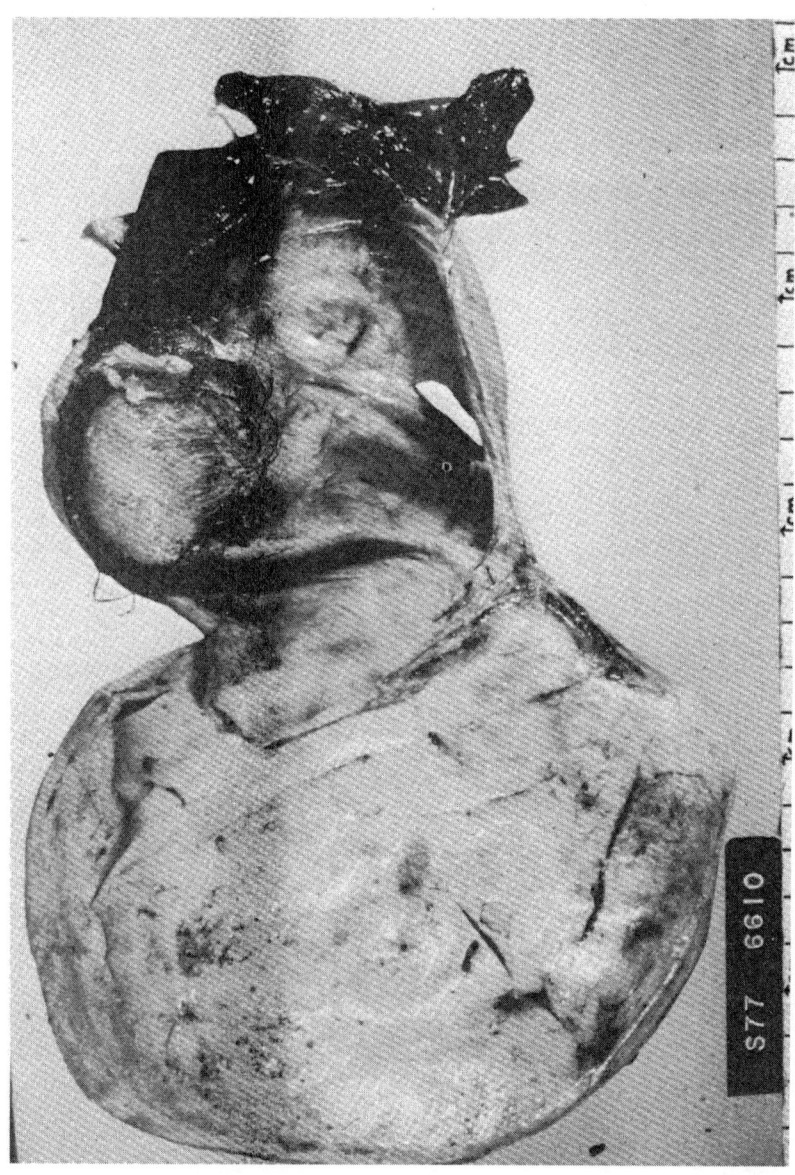

FIGURE 8. Benign cystic teratoma with fibroma. (Pennsylvania Hospital 77-6610.)

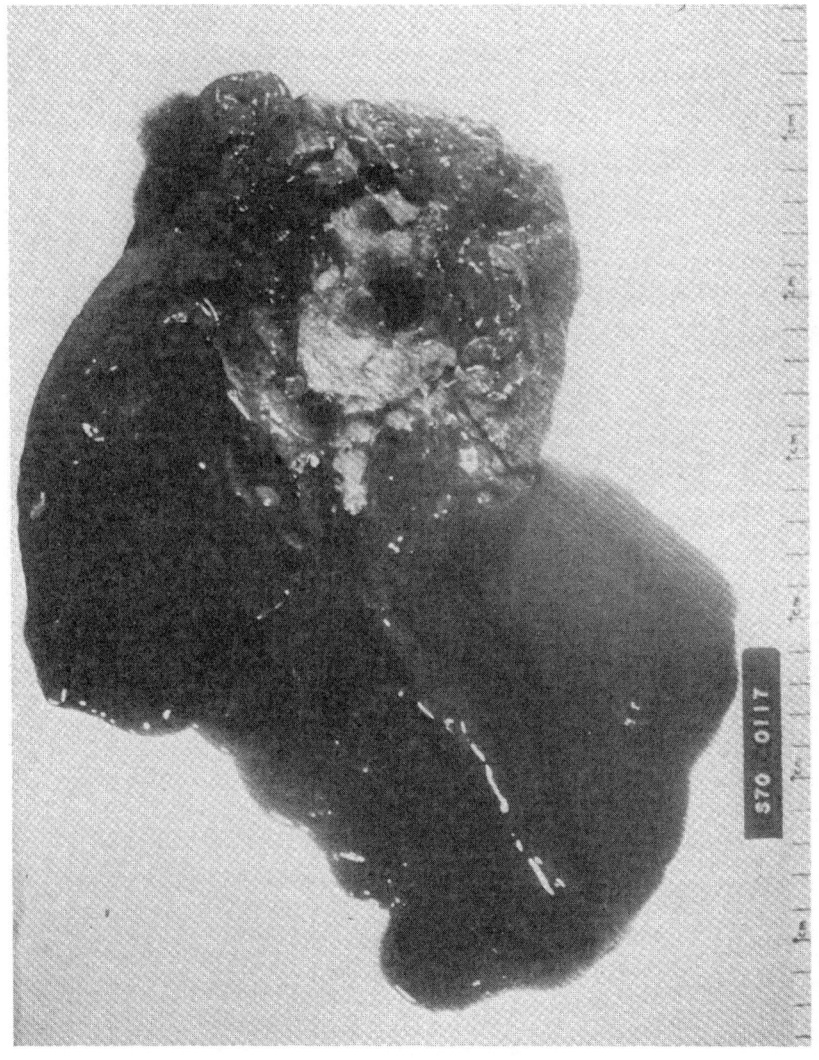

FIGURE 9. Benign cystic teratoma with mucinous cystadenoma. (Pennsylvania Hospital 70-117.)

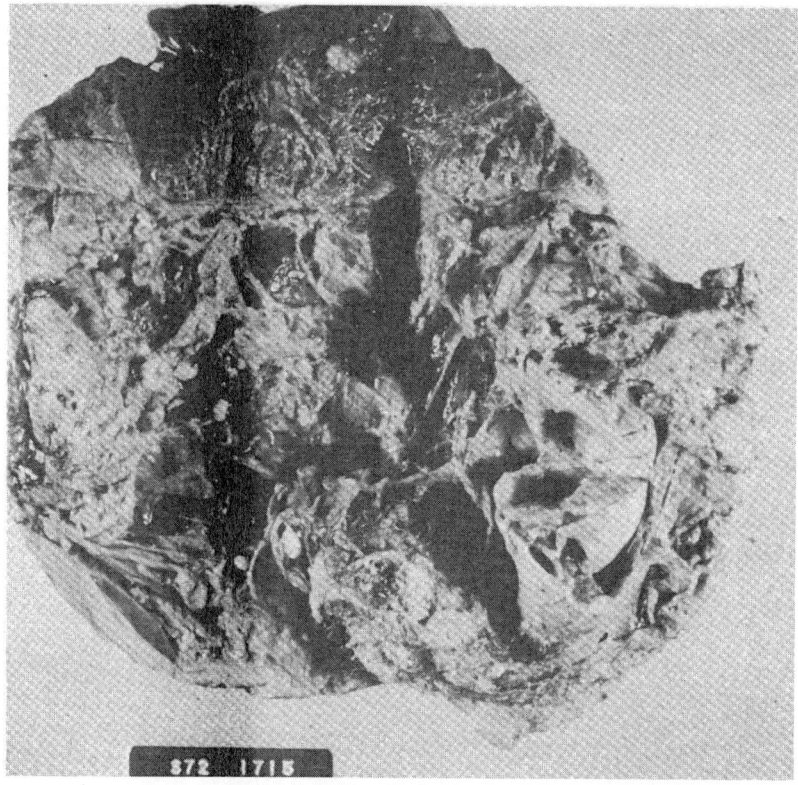

FIGURE 10. Cystic teratoma with adenocarcinoma. Sectioned surface shows solid areas and hair. (Pennsylvania Hospital 72-1715.)

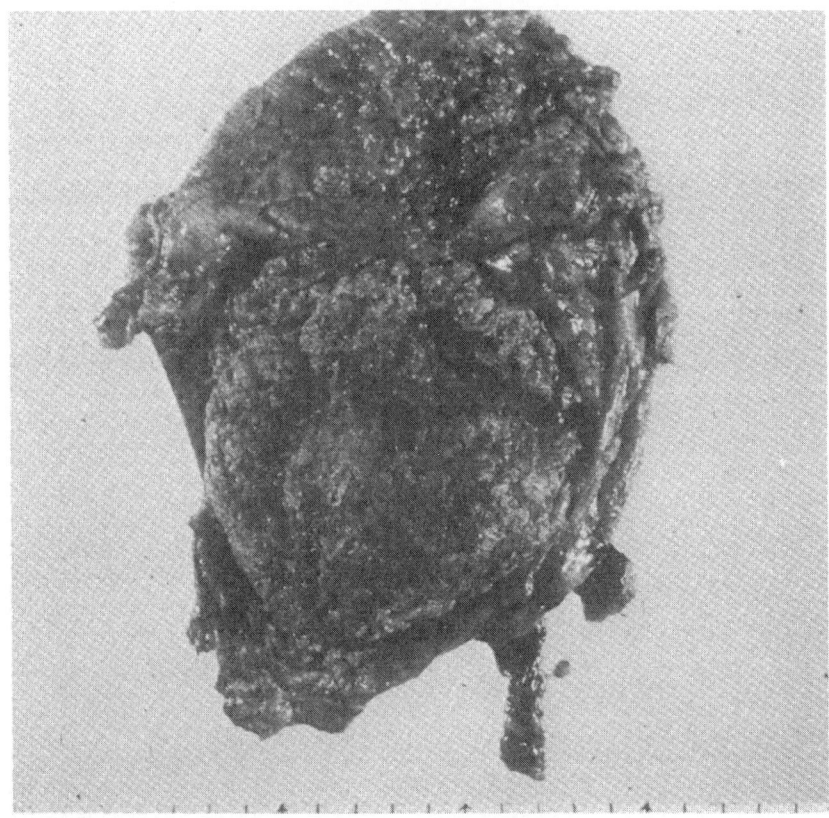

FIGURE 11. Benign cystic teratoma with epidermoid carcinoma (solid areas). (Pennsylvania Hospital 66-1219.)

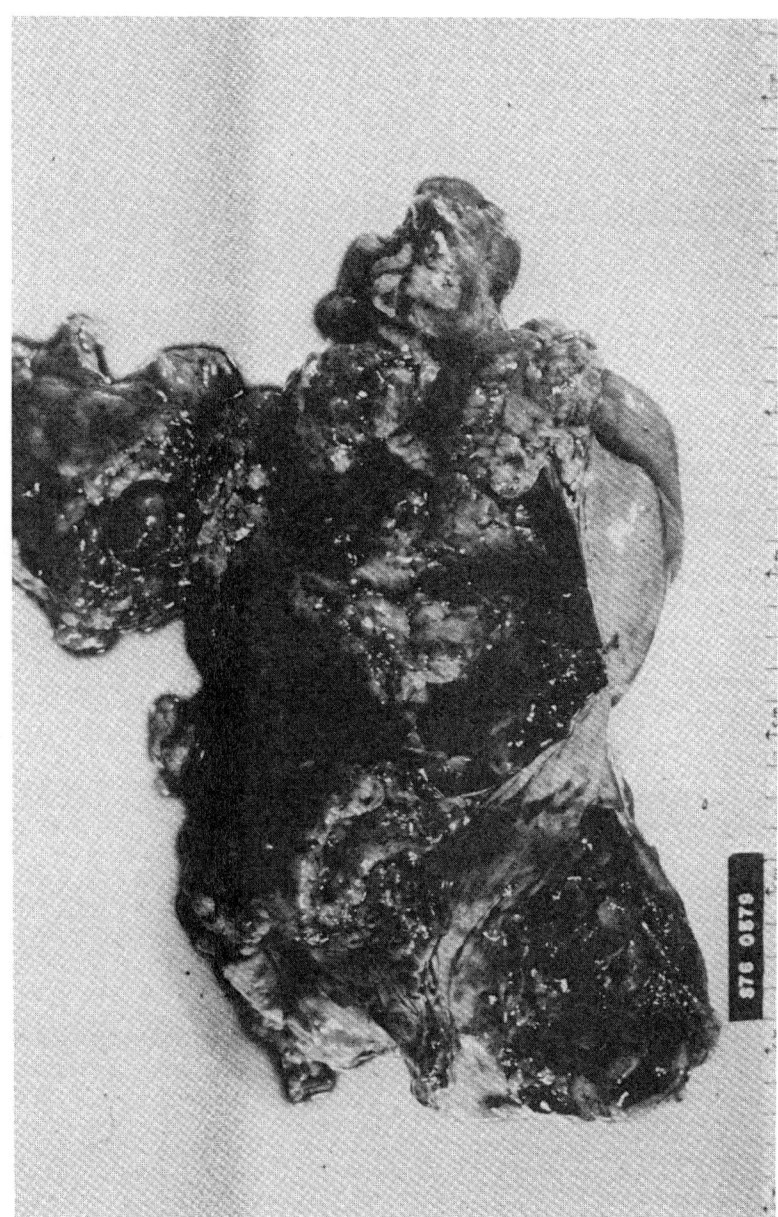

FIGURE 12. Solid teratoma. Predominantly solid tumor with small cystic areas. (Pennsylvania Hospital 76-0579.)

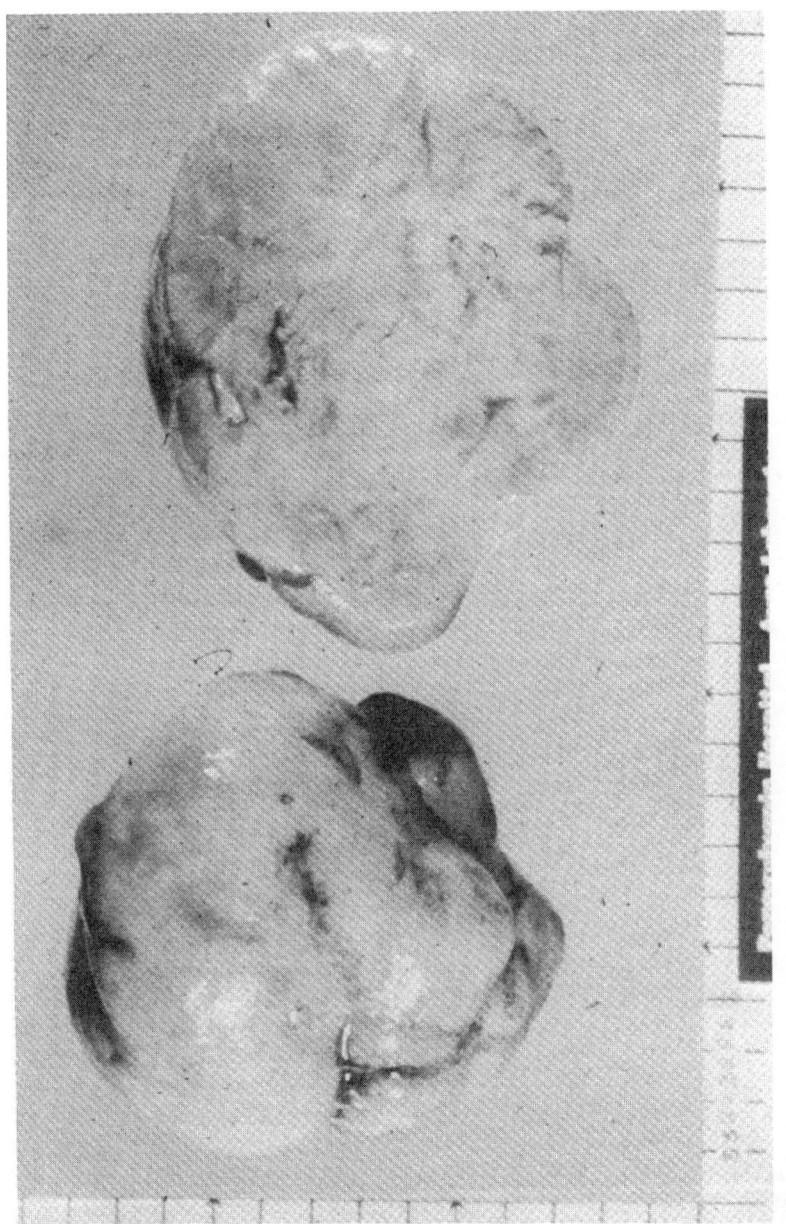

FIGURE 13. Dysgerminoma—external and sectioned surface. Note characteristic bosselated exterior and lobulated sectioned surface. (Pennsylvania Hospital 58-3286.)

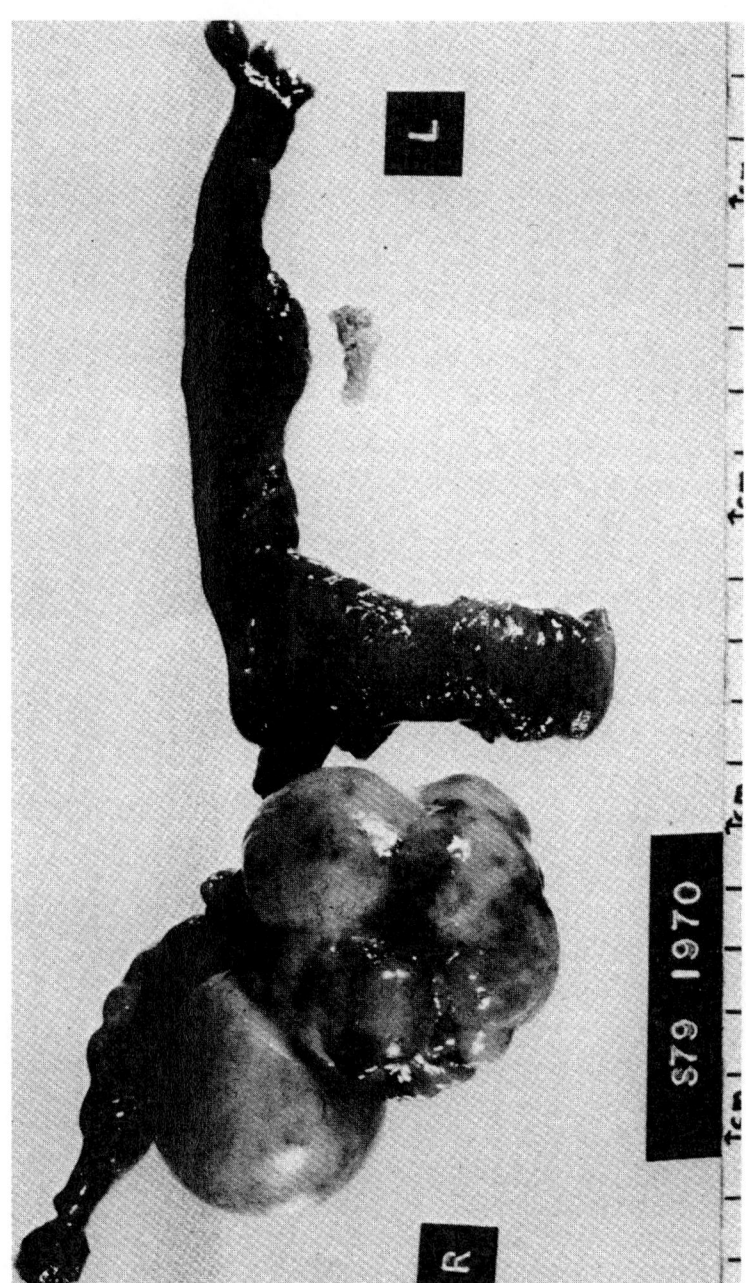

FIGURE 14. Dysgerminoma in XX/XO patient. Streak gonad on left. (Pennsylvania Hospital 79-1970.)

REFERENCES

1. Lee PA: Normal ages of pubertal events among American males and females. *J Adolescent Health Care* 1:26–29, 1980

1a. *1972 Facts and Figures*, New York, American Cancer Society, 1971

2. Willis RA: *The Pathology of Tumors*. New York, Mosby, 1948

3. Barber HRK, Graber EA: Gynecological tumors in childhood and adolescence. *Obstet Gynecol Survey* 28:357–381, 1973

4. McDonough PG: The adolescent gynecologic patient and her problems. *Clin Obstet Gynecol* 22:491–507, June 1979

5. Hobbs JE: Sweat gland tumors. *Clin Obstet Gynecol* 8:946–951, 1965

6. Apfelberg DB, Morton RM, Lash H, et al: Progress report on extended clinical use of the Argon laser for cutaneous lesions. *Lasers Surg Med* 1:71–83, 1980

7. Birch HR, Senday DR: Granular cell myoblastoma of vulva. *Obstet Gynecol* 18:443–453, 1961

8. Hahn GA: The carbon dioxide laser in gynecology. *Phila Med* 74:344–347, 1978

9. Herbst AL, Scully RE: Adenocarcinoma of the vagina in adolescence. *Cancer* 25:745–757, 1970

10. Ober WB, Edgecomb JH: Sarcoma botryoides in the female urogenital tract. *Cancer* 7:75–91, 1954

11. Hilgers RD, Malkasian GD, Doule EH: Embryonic rhabdomyosarcoma of vagina. *Am J Obstet Gynecol* 107:485–502, 1970

12. Beral J: Cancer of the cervix: A sexually transmitted infection? *Lancet* 1:1037–1040, 1974

13. Stafl A, Mattingly RD: Colposcopic diagnosis of cervical neoplasia. *Obstet Gynecol* 41:168–176, 1973

14. Baggish MS: HIgh power density carbon dioxide laser therapy for early cervical neoplasia. *Am J Obstet Gynecol* 136:117–125, 1980

15. Dorsey JH, Diggs ES: Microsurgical conization of the cervix by carbon dioxide laser. *Obstet Gynecol* 54:565–570, 1979

16. Fraser IS, Baird DT: Endometrial cystic hyperplasia in adolescent girls. *J Obstet Gynaecol Br Cwlth* 79:1009–1015, 1972

17. Goldstein DP: Preferable management of gestational trophoblastic disease: Benign and malignant, in Reid, DE, Christian, CD (eds): *Controversy in Obstetrics and Gynecology*. Philadelphia, Saunders, pp 219–234, 1974

18. Scully RE: Recent progress in ovarian cancer. *Human Pathol* 1:73, 1970

19. Sternberg WH: Nonfunctioning ovarian neoplasma. *The Ovary*, International Academy of Pathology Monograph. Baltimore, Williams & Wilkins, 1963, pp 209–254

20. Abell MR, Holtz F: Ovarian neoplasma in childhood and adolescence. *Am J Obstet Gynecol* 93:850–866, 1965

21. Morris J, McLean O, Scully RE: *Endocrine Pathology of the Ovary*. St. Louis, Mosby, 1968, pp 65–95

22. Whelton JA, Christian HJ: Long term survival following conservative surgery for bilateral arrhenoblastoma. *Obstet Gynecol* 27:210–213, 1966

23. Stenwig JT, Hazekamp JT, Beecham JB: Grannulosa cell tumors of the ovary. A clinicopathological study of 118 cases with long-term follow-up. *Gynecol Oncol* 7:136–152, 1979

24. Peterson WF, Prevost EC, Edmunds FT, et al: Benign cystic teratomas of the ovary. *Am J Obstet Gynecol* 70:368–382, 1955

25. Teilum G: Tumors of germinal origin, in *Ovarian Cancer*, International Union against Cancer Monograph Series 11. Berlin, Heidelberg/New York, Springer-Verlag, 1948, p 58

26. Karlen JR, Kastelic JE: Endodermal sinus tumor of the ovary: An improving prognosis. *Gynecol Oncol* 10:206–216, 1980
27. Novak ER, Woodruff JD: *Gynecological and Obstetrical Pathology*, Philadelphia, Saunders, 1979, pp 476–503
28. Scheelhaus HF, Trujillo JM, Rutledge FM, et al: Germ cell tumors associated with XY gonadal dysgenesis. *Am J Obstet Gynecol* 109:1197–1204, 1971

CHAPTER 10

Laboratory Evaluation in Adolescence

Alfred M. Bongiovanni

1. CLINICAL EVALUATION

The evaluation of the girl approaching the age of puberty on into adolescence does not usually require extensive laboratory investigation. Often, the status may be determined by a careful history, physician examination, and periodic assessment. It is important to examine the pattern of physical growth through the first decade of life in order to rule out preexisting conditions that may contribute to aberrations of further development. For example, extremely slow prepubertal linear growth will suggest gonadal dysgenesis or various degrees of hypopituitarism. Unduly diminished weight gain may be the result of poor nutrition, which in itself can delay puberty as can impending anorexia nervosa or chronic systemic debilitating disease.

The history should consider the socioeconomic condition of the family, insight into the psychological milieu, and the pattern of pubertal development within the family with special reference to the age of menarche in the mother and other female members.

The particular features of the physical examination that herald the normal onset and progress of puberty in the girl includes the typical growth spurt, the appearance of pubic hair, breast development, and genital changes.[1,2] The increase in the velocity of linear growth characteristic of early puberty generally begins at about 11 years of age and reaches its

Alfred M. Bongiovanni • Departments of Pediatrics and Pediatrics in Obstetrics and Gynecology, School of Medicine, University of Pennsylvania, Philadelphia, Pennsylvania 19104; Department of Perinatal Endocrinology, Pennsylvania Hospital, Philadelphia, Pennsylvania 19107; and Department of Pediatrics, Thomas Jefferson University, Philadelphia, Pennsylvania 19107.

peak at 12 years. Thereafter, there is a decrease in the velocity with the onset of menarche shortly before the cessation. The development of pubic hair has been divided into five stages (Fig. 1). At Stage I, there is no true pubic hair other than the vellus. At Stage II, there is a sparse appearance of slightly pigmented hair confined to the labia majora. At Stage III, the quantity of hair increases somewhat and becomes darker and coarser with slight spread over the pubic junction. At Stage IV, the amount and appearance is approaching that of the adult but smaller in amount and without spread to the medial surface of the thighs. Finally at Stage V, there is the complete adult pattern with spread to the medial surface of the thighs in the form of an inverted triangle of the typical female pattern (see Fig. 1).

The development of the breasts is also divided into five stages (Fig. 2). In the beginning, Stage I, there is only prominence of the papillae. Stage II is characterized by elevation of the breast and papillae presenting a small mound which is confined to the areolar diameter. At Stage III, there is a further enlargement of all tissue, but without separation of the contours of the breasts and the areolae. At Stage IV, there is a definite projection of the areolae and papillae with the appearance of a secondary prominence of these structures above the level of the breast. At Stage V, wherein there is full development, the breasts have reached their full size (which is variable) and with recession of the areolae into the general structure of the breast itself so that only the papillae project.

The prepubertal glistening pink surface of the vaginal mucosa becomes dull and reveals some small amount of secretion, which becomes

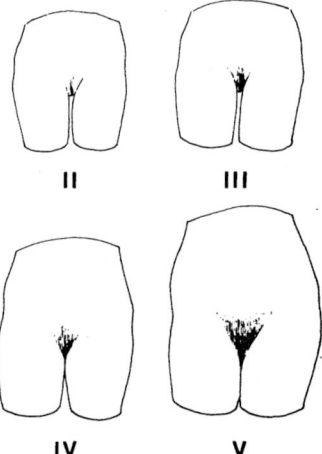

FIGURE 1. Stages of pubic hair development in females modified after Marshall and Tanner.[1] In Stage I, there is no pubic hair. In Stage II, there is a small amount limited to the labia with gradual spread through the ensuing stages over the pubes and slightly to the medial thighs.

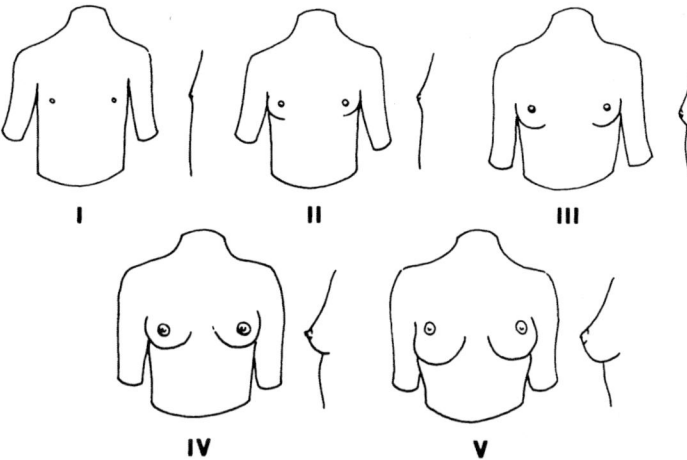

FIGURE 2. Stages of breast development in the female modified after Marshall and Tanner.[1] These are shown both from the anterior and lateral views. At Stage II, there is some fullness of the breasts with early prominence of the nipples, which gradually increase. At Stage IV, the nipples actually protrude but by Stage V they are absorbed into the general breast tissue.

a whitish discharge just prior to menarche. The labia minora, which in the beginning are clearly perceptible only at their juncture at the fourchette, enlarge and gradually surround the introitus as a more definitive structure. The average age of menarche is 13 years, but the normal range extends from $9\frac{1}{2}$ to 15 years. The earliest signs of breast development or of the appearance of pubic hair are also variable and may depart by ± 2 years from the ages described. Such variations in themselves are not cause for intensive laboratory investigation. In particular, the delay of menstruation in the face of retarded but clearly progressive development of breasts, genitalia, pubic hair, and linear growth are not cause for immediate study. However, the data provided in Tables I–III may be employed for the assessment of the hormonal progression when there is concern. It should be emphasized that these data, which are classified according to chronologic age, correlate better with the epiphyseal maturation. Therefore, it would be better to employ the bone age as the standard of comparison. It must also be emphasized that these tables represent a compilation of data from a large number of laboratories, and it is important to employ the standards of the particular laboratory providing the services.

The study of the vaginal cytology remains a helpful and simple guide, which can be carried out at the time of the examination. Specimens are obtained from the lateral wall of the posterior vagina obtained by aspiration with a pipett or collected on a cotton applicator. While this procedure may be innocuous, it is sometimes distressing to the young girl. The total absence of any estrogen is manifested by the presence of almost entirely small parabasal cyanophilic cells. On the other hand, the presence of 20% or more large squamous eosinophilic cells is good evidence that the hypothalamic–pituitary–ovarian axis has been activated. This test alone may serve to temporarily assure the onset of puberty when it has been delayed or contrarily suggest the need for further study when there are no signs of puberty and exclusively basal cells are present in the smear.

Prior to the increased velocity of linear growth, there is an increase in body mass, which begins at around 7 years in females, and is accompanied by a gradual increase of body fat. One controversial theory suggests a critical amount of body fat and a mean weight of 48 kg is required for the occurrence of menarche. While there may be some question concerning the immutability of this factor, it is true that malnutrition is accompanied by delayed menarche, whereas the reverse is true in the moderately obese (see Chapter 2).

2. PITUITARY HORMONES

2.1. Gonadotropins

Normal levels of luteinizing hormone (LH) and follicle-stimulating hormone (FSH) in the serum[3–6] at various ages is shown in Fig. 3 and Table I. However, as emphasized in the other sections of this chapter, it may be more appropriate to employ the bone age as the standard of reference. There are two problems in the evaluation of the results of these

TABLE I. Pituitary and Gonadal Hormones[a]

Pubertal stage	FSH (mIU/ml)	LH	Estradiol (pg/ml)	Estrone (pg/ml)	Progesterone (ng/ml)	Prolactin (ng/ml)
I	2–4	1–6	8.2 (1–20)	13 (4–29)	0.28 ± 0.04	3–18
II	3–10	1–9	16.4 (10–24)	21 (10–33)	0.76 ± 0.10	
III	3–15	3–15	25 (7–60)	30 (15–43)	0.85 ± 0.10	
IV	4–20	4–20	47 (21–85)	36 (16–77)	1.13 ± 0.11	
V	4–20	5–25	111 (34–170)	61 (29–77)	1.51 ± 0.16	
Adult						
Follicular	5–20	5–25	24–68		0.1–1.0	4–21
Luteal	8–25	20–90	20–40		3.0–15.0	4–21

[a] Data from Ref. (3–6, 10).

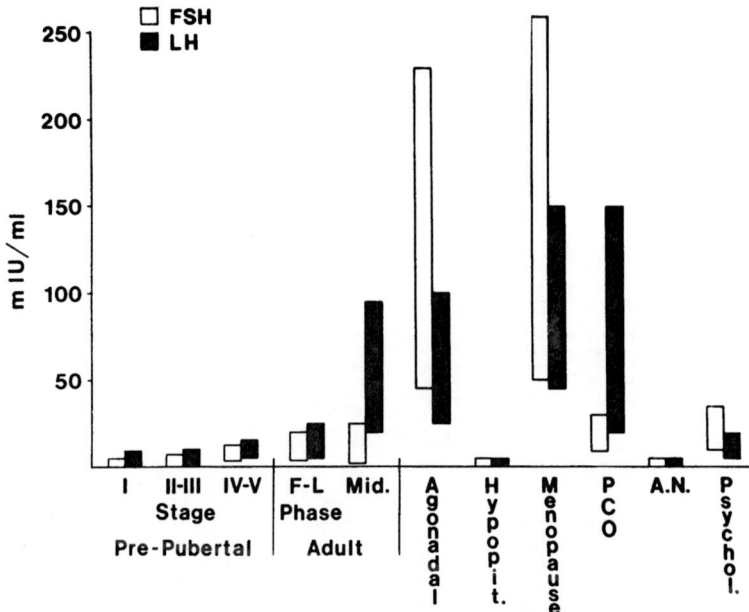

FIGURE 3. Plasma level of FSH and LH in females at the various stages of sexual development on into adulthood. In the adult the values are shown for the F–L (early follicular and late luteal) phase and for the midcycle. Included are the levels in various disorders. PCO, Polycystic ovarian disease; A.N., anorexia nervosa; Psychol., various forms of hypothalamic menstrual disorders.

tests before and during adolescence. Prior to the appearance of any secondary sexual changes, or even in the early stages of puberty, the levels of these hormones may be extremely low and therefore of little assistance in arriving at the diagnosis of insufficiency. The other problem is that these hormones are normally secreted episodically, and this is especially true of LH. Therefore, one cannot be certain that the specimen has been drawn at a time that a significant level will have been achieved. However, in very early puberty, there is an enhanced secretion of LH that occurs during sleep (Fig. 4). In cases of doubt it may be advisable to obtain two or three specimens during the first 90 min of sleep by maintaining a continuous intravenous infusion of saline in order to facilitate the withdrawl of specimens without disturbing the subject.

In the face of low levels of gonadotropins, the pituitary integrity with regard to these hormones may be tested by the administration of the gonadotropin-releasing factor (GnRF).[7,8] It must be noted, however, that the elevation of the serum gonadotropin following such stimulation is slight prior to puberty and increases markedly during the peripubertal period, becoming even greater in adult life (Fig. 5). In girls, this is particularly true of the LH. Thirty minutes following the intravenous administration of 100 μg of GnRF, the level of LH may rise from a level

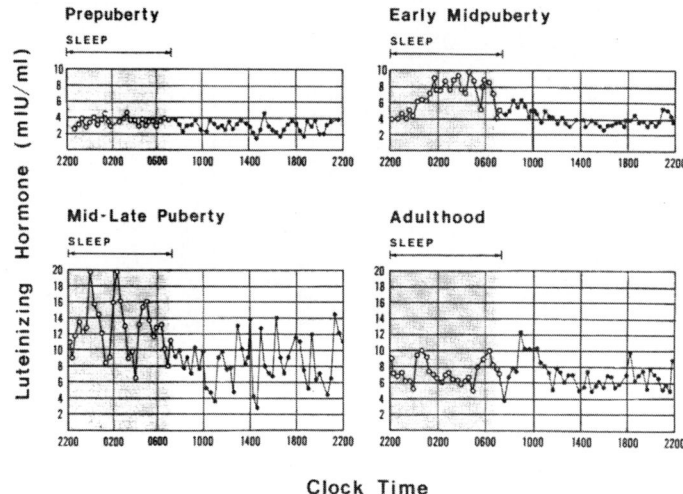

FIGURE 4. Plasma levels of LH during sleep–wake periods in the various stages of sexual development of the female. Note the increased amplitude and frequency during early midpuberty and late puberty. (From Ref. (18). Reproduced with the permission of the publisher, the American Fertility Society.)

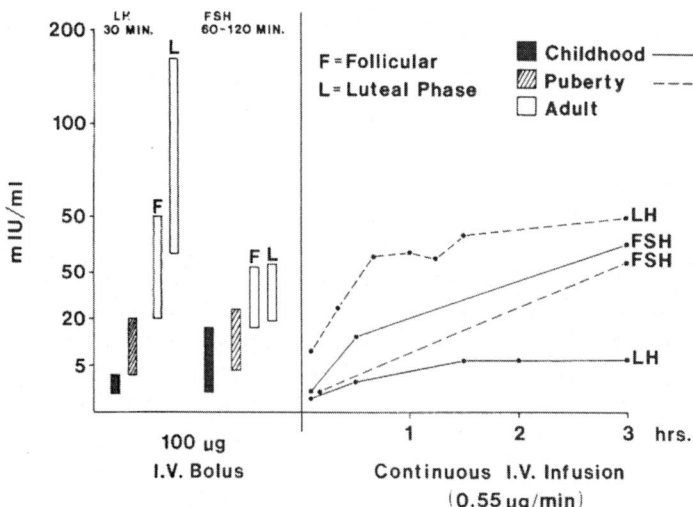

FIGURE 5. Responses of the plasma level of LH and FSH to the intravenous administration of GnRh. The responses to a single bolus are shown on the left; in the adult, the levels are shown for the follicular as well as the luteal phase. On the right are shown the responses to a continuous infusion of GnRh.

slightly below 1 mIU to only twice the baseline value. With the appearance of the secondary sexual characteristics, baseline levels of approximately 5–10 mIU/ml will double or more in value. Following completion of sexual development, the rise will be even more marked and will vary with the stage of the cycle. The rises in response to such stimulation in the late follicular phase will achieve values of 50 mIU/ml LH or greater. At this later age, the LH response to continuous infusion is biphasic (Fig. 5), suggesting two pools in the pituitary, one released early and the other possibly newly synthesized hormone. In the female (but not the male), the response of the FSH is significant even prior to puberty provided the pituitary gland is responsive. The relatively low levels of early puberty will at least double and provide evidence of the integrity of the FSH. In addition, primary ovarian deficiency may reveal itself even at an early stage by an extremely sharp rise in the FSH to such stimulation. Although not employed as a routine test, the repetitive pulsatile administration of the releasing factor will show gradually increasing degrees of response. Since estrogens are thought to sensitize the pituitary to the action of GnRF, it is presumed that this effect is estrogen dependent and a reflection of increased production of this hormone by the ovary during the preceding stimulation.

A number of other diagnostic features of these tests have been described in the appropriate sections of this book. Among specific examples is the relatively greater rise of LH after GnRF in the polycystic ovarian syndrome. Indeed the tonic preponderance of LH in this disorder has been well described. On the other hand, an irregularity in the baseline relationship of FSH to LH (a higher ratio) has been described in certain forms of psychogenic hypomenorrhea.

Disorders of the hypothalamus wherein GnRF is deficient have been described in both males and females. At the present time, methods for the determination of serum levels of GnRF are not available, and it is therefore not possible to make such a diagnosis directly. The virtual absence of FSH and LH in the serum followed by an appropriate response to GnRF affords indirect proof of this condition. However, it should be noted that the response to a single stimulation is apt to be subnormal in this situation. The repetition of the stimulation at intervals of 2–3 hr will usually reveal a normal response. This is presumably because of the need to prime the pituitary, which has not been previously exposed to GnRF probably by way of the release of ovarian estrogens during the earlier tests.

The measurement of the gonadotropins are of great value in the assessment of several problems in adolescence. In gonadal dysgenesis they are elevated, especially the FSH. In this condition, there is an increase during the first 2–3 years of life. Thereafter they fall to the normal levels for the age and rise markedly around the time of puberty when they remain elevated thereafter. Since the polycystic ovarian syndrome begins with the early menarche, the typical elevation of LH relative to the FSH is present. In hypothalamic (psychological) amenorrhea the reverse is true, i.e., a relatively higher FSH. In hypopituitarism, whether total or limited to failure of the gonadotropins, there is an absence of LH or FSH or both, and there is no response to the administration of GnRF. Other conditions associated with variations in gonadotrophin levels are illustrated in Fig. 3.

2.2. Prolactin

An increasing number of cases of primary amenorrhea are being associated with prolactinoma.[9] Therefore, in some clinics it has become routine to obtain a measurement of the serum prolactin when the menarche is unduly delayed. Galactorrhea is not necessarily a feature, and secondary sexual development may be normal. The normal levels in prepubertal girls is 3–18 ng/ml increasing to 4–21 ng/ml in adulthood. While elevated levels in the presence of delayed adolescent development should be an invitation to full-scale investigation of the pituitary by such methods as polytomography, it must be mentioned that certain other conditions are associated with an increase. These include primary hypothyroidism, oral contraceptives, breast stimulation, herpes zoster, and psychotrophic and antihypertensive drugs.

3. OVARIAN HORMONES

In Table I there are shown the serum levels of estradiol and progesterone at various ages in the female.[3,10] They both show gradual rises beginning just shortly before pubertal development. They may be employed as a measure of ovarian function. In the face of ovarian malfunction, these may be expected to be low and will be accompanied by rises in the gonadotrophins. Although it is possible to test ovarian response to the administration of gonadotrophin, especially with regard to the levels of estradiol, this test is not really necessary. The cyclic changes in these hormones are seen in adulthood, and with ovulation the progesterone reaches values in excess of 10,000 ng%. However, during the early years of menstruation when ovulation frequently does not occur, these marked elevations will not be observed. The changes in serum estradiol and progesterone are of great value in the management of infertility, a problem which does not present itself until later in life. The urine levels of pregnanediol (Table II) has long been employed as a measure of progesterone secretion and is a reflection of the blood levels of progesterone. It is used less frequently but remains a test of some value in special circumstances. Urinary estrogens are no longer employed in the evaluation of the en-

TABLE II. Levels of Androgenic Hormones in Serum of Females[a]

Pubertal stage	Chronologic age (years)	DHA sulfate (μg/dl)	Free DHA (ng/dl)	Testosterone		DHT (ng/dl)	A'dione (ng/dl)	17-OHP (ng/dl)
				Total (ng/dl)	Free (pg/ml)			
I	6–9	15 ± 12.4	104 ± 38	21 ± 2	1.1	0.06	6.9	14.5 ± 0.03
II	8–11	43 ± 8.8	173 ± 68	24 ± 3		0.10		20 ± 0.05
III	10–12	49 ± 22	329 ± 60	35 ± 3		0.14	8.4	28 ± 0.03
IV	11–14	71 ± 54	427 ± 161	37 ± 3		0.15		36 ± 0.05
V	12–15	86 ± 48	491 ± 120	51 ± 3		0.16		29 ± 0.04
	13–17	120 ± 42	634	57 ± 3	4.3 ± 1.2	0.19	55–220	37 ± 0.05
Follicular								30–100
Luteal								150–200

[a] Data from Refs. (11–13). DHA, Dehydroepiandrosterone; DHT, dihydrotestosterone; A'dione, androstenedione; 17-OHP, 17-hydroxyprogesterone.

docrine status in the adolescent, although they do remain of value in monitoring pregnancy.

4. ANDROGENS

4.1. In Serum

The major androgenic hormones to be found in the female serum are testosterone (T), androstenedione (A), and dehydroepiandrosterone (DHEA). A large portion of the T is present in the serum tightly bound to a specific β globulin and a small amount (approximately 1%) is "free." DHEA is present for the most part in the circulation as its sulfate (DHEAS), and while a very small amount is unconjugated (free) this amount does not seem to have any particular value in diagnosis. The serum levels of these androgens are given in Table II.[11-14]

The A present in the serum of women is secreted as such in equal amounts by both the ovaries and the adrenals except in the midcycle, at which time about 65% of the serum level arises from the ovaries. Very little of the A represents significant conversion in the periphery from other hormones, such as T or DHEA. About 10–20% of the serum T arises from the ovaries and 5–30% from the adrenals. Some 50% of the T measured in the blood is not a reflection of direct secretion but rather represents the peripheral conversion of A to T. More than 90% of the DHEAS measured in the blood arises from the adrenals. Therefore, of the various androgens to be studied in the female, the last is perhaps the least equivocal as concerns its origin and it usually reliably reflects adrenal contributions.

Quantitatively, DHEAS is the most abundant, but not the most potent, of the androgens generally measured in the laboratory. Its level during childhood are lower and gradually increase through puberty into adolescence, as shown in Table II. It is believed to be the hormone most responsible for the appearance of pubic hair, the slight clitoral enlargement seen at puberty, and the changes in the labia majora. Therefore, in the condition known as premature adrenarche, wherein there is the early development of pubic hair without other changes, the serum DHEAS will be moderately elevated in the presence of normal quantities of the pituitary and gonadal hormone for the chronologic age. There are some differences in the data on this last point so that a few investigators have also reported slight but significant rises in the serum estradiol as well. Elevations of the serum DHEAS well above the normal for any age strongly implicate adrenal disease. Extraordinarily high levels in excess of 1000 μg% suggest an adrenal tumor. Although it is possible to measure the free unconjugated DHEA, this merely adds to the expense of laboratory investigations as previously noted, and as of this writing it is not known to be of any specific diagnostic significance. In many forms of significant hirsutism in adolescence, which may indeed be an early earmark of polycystic ovarian syndrome, DHEAS has been reported to be significantly but not greatly elevated and is thought in some quarters to be an invitation for corticosteroid treatment. In the adolescent girl with significant hirsutism and menstrual irregularity, it has become customary to measure the serum levels of T, A, and DHEAS. However, there is considerable evidence that the serum level of T, and in particular the free component, is often elevated, whereas the other androgens may be normal.[11] The matter of 17-hydroxyprogesterone will be discussed.

The interpretation of the levels of A and T remain troublesome and controversial. This is because of their dual origins, adrenal or ovary, and the knowledge that some T does arises from peripheral conversion of other hormones. With regard to T, the free level is more meaningful because it is the biologically active form, and the total level includes the largely biologically unavailable form. In some studies, many hirsute girls have elevations of both A and T. Some investigators have determined that the source is usually ovarian.[15] Such findings are consistent with the polycystic ovarian syndrome, which is a heterogeneous disorder.

4.2. Urinary Hormones

Measurement of the urinary 17-ketosteroids (Table III) has limited value in the assessment of endocrine problems in adolescence. They are expected to be elevated in the various virilizing conditions but may be normal in the face of elevated serum androgens previously discussed. They are in great measure a reflection of adrenal function. The elevation of serum androgens described in the various masculinizing syndromes, while significant, are quantitatively small in terms of the excretion of

TABLE III. Urinary Steroid Hormones

	Stages					
	I	II–III	IV	Adult	Follicular	Luteal
17-Ketosteroids (mg/day)	1–3	2–5	3–6	5–11		
Corticoids (Porter–Silber) (mg/day)	2–3	3–5	4–6	5–9		
Free cortisol (μg/day)	10–30	15–40	20–60	30–100		
Pregnanetriol (mg/day)	0.5–0.8	0.5–1.5	1.0–1.5	2.5–4.5	1.5–2.5	2.5–4.5
Pregnanediol (mg/day)	0.1	0.2	0.4–1.5	2–10	1.5–2.5	3.5–10.0

urinary metabolites expressed as 17-ketosteroids. In classical adrenal hyperplasia or most adrenal tumors, the 17-ketosteroids are significantly elevated. In the former, the urinary pregnanetriol will also be elevated. These latter changes suggest the need for adrenal suppression with corticosteroids.

Although the urinary gonadotropins continue to be used in some quarters, they are less often employed in general. However, they have the advantage of representing integrated values, whereas serum levels, especially of LH, are pulsatile.

4.3. Suppression Tests

Differential suppression tests have been employed to distinguish the ovarian from the adrenal course of elevated androgens from which a program of treatment is designed. It is recognized that such an approach is not universally accepted among the experts, but this type of testing appears to have merit in the elucidation of many hyperandrogenism cases especially when the results are clear-cut. In order to suppress the ovarian contribution, a combination-type oral contraceptive may be given such as 2.0 mg Ortho-Novum® bid for 3 weeks. Suppression of the ovary is relatively slow, and the second specimen of blood should not be obtained until 3 weeks after the initiation of treatment in order to determine whether the A and/or T have been suppressed. To test for the adrenal contribution, 5.0 mg of prednisone may be given twice daily for 5 days. On the fifth day, a significant diminution of A and/or T should be noted in order to incriminate the adrenal as the source. When one of these tests is clearly successful but not the other, it may be rationally concluded that the responsible gland has been identified, although as noted, this matter remains controversial. It is probably that in many instances both glands are contributing to the problem.

5. ADRENAL FUNCTION

Both adrenal insufficiency and hypercortisolism (Cushing syndrome) may be associated with disturbances in general development and sexual maturation. In primary adrenal insufficiency, the best single test is the early morning determination of the plasma cortisol, at which time of the day it is at its peak. Levels consistently below 5 μg% strongly suggest the diagnosis. The adrenal reserve may then be tested by the intravenous administration of a bolus of 250 μg synthetic adrenocortocotrophin (Cortrosyn). This should evoke at 1 hr an increase to 15 μg% or greater. The complete failure of response is adequate proof of adrenal insufficiency.

Cushing syndrome, on the other hand, will reveal consistently high early morning and late afternoon plasma levels of cortisol without any diurnal variation and usually are above 15 μg% at both times. In addition, the total urinary corticoids are generally elevated well above the normal values. Perhaps the most valuable test is the measure of free urinary cortisol. When this is well over 100 μg/30 kg body wt/day, this is further evidence of Cushing syndrome (Table III). The urinary 17-ketosteroids may also be elevated.

Some forms of late onset congenital adrenal hyperplasia may account for hirsutism and irregular menses in early adolescence. The most valuable test to detect these milder forms of the disease due to 21-hydroxylase deficiency at the present time is the measurement of the blood 17-hydroxyprogesterone[16,17] (Table II). Levels above 400 ng% are strongly suggestive of this condition and virtually diagnostic if they exceed 1000 ng%. It must be remembered that the levels of this compound are elevated during the luteal phase of the cycle. In some patients the baseline is equivocal, and the picture can be brought into better focus by stimulation of the adrenal. This challenge is generally best performed after a previous midnight dose of 1.0 mg Decadron®. If in the next morning within 30–60 min of the administration of 250 μg Cortrosyn the blood level of 17-hydroxyprogesterone exceeds 1000 ng%, this is strong evidence for 21-hydroxylase deficiency.

In order to distinguish Cushing's disease (usually the result of a pituitary adenoma) from adrenal tumor a second "high dose" suppression test is in order. Decadron, 2.0 mg, is given every 6 hr for 2 days. On the second day, the plasma cortisol would show at least a 50% suppression with Cushing's disease, but in general the adrenal tumor will not respond.

There is appearing with increasing frequency a similar attenuated form of congenital adrenal hyperplasia due to 3-β-hydroxysteroid dehydrogenase (HSD) deficiency.[17] With this disorder the principal plasma steroid of interest is 17α, 3-dihydroxy-5-pregnen-20-one (17-OH-pregnenolone). Normal baseline levels from limited data indicate values of 80–150 ng%. A ratio of 17-OH-pregnenolone to 17-hydroxyprogesterone that is greater than 2.8 also suggests this particular condition. With stimulation of the adrenal gland using Cortrosyn, as already mentioned, this ratio becomes exaggerated and is about 10, and the levels of 17-OH-pregnenolone are greater than 800 ng%. Both of these conditions of late onset congenital adrenal hyperplasia respond well to replacement doses of corticoids.

6. SEX CHROMOSOME ANALYSIS

All girls of less than normal stature for age and with absence of secondary sexual development should be investigated for possible gonadal dysgenesis. The simplest and least expensive test is the buccal smear. While the technique is not difficult, it does require experience on the part of the laboratory for a correct preparation and interpretation. The presence of the sex chromatin near the nuclear membrane (Barr body) is to be seen in 30% or more of the cells examined. A low or indeterminate count should be occasion to determine the karyotype. Properly conducted, this will better reveal XO dysgenesis and various forms of mosaicism. In addition, the karyotype will also reveal the occasional case due to an isochromosome of the long arm of the X. This latter condition may also be suspected if the Barr bodies are large or apparently high in number.

7. THYROID TESTS

Deficiency of the thyroid hormone is usually associated with a delay in general somatic development and often, but not always, in the appearance of the secondary sexual characteristics. In severe deficiency, the general physical examination including the expected suboptimal rate of growth should lead to suspicion. At the present time the most reliable single test is the serum thyroxine (T4) level. The normal range, which varies slightly between laboratories, is 5–12 μg%. The levels at the time of puberty tend to be somewhat higher than earlier or later so that results within the lower normal range should be regarded with suspicion. In these circumstances, the thyroid-stimulating hormone (TSH) should be determined. The normal values for adolescent females lies between 7 and 12 μU/ml. In thyroid deficiency, levels of TSH are extremely elevated. It should be mentioned that the normal levels of TSH for younger girls is considerably lower varying from 1 to 6 μU/ml. An additional confirmatory test that may be ordered at the time of the T4 is the T3 uptake, which is an indirect measurement of the available thyroxine-binding globulin and is *not* a measure of the serum triiodothyronine (T3). It also is low in hypothyroidism. It is likely that these few tests will settle the diagnosis, but in the rare event that the situation remains equivocal, the thyrotrophin-releasing hormone (TRH) may be administered to further clarify the issue.

Following an overnight fast, a baseline specimen is drawn and 500 μg of TRH is injected by vein over 30 sec. Blood specimens are obtained at 30, 45, and 60 min. In euthyroid individual, the rise will occur with a peak between 30 and 45 min not exceeding 25 μU/ml, whereas in hypothyroidism, the rise will exceed 200 μU usually having started from an already elevated baseline. Furthermore, the elevation in hypothyroidism is extremely prolonged and does not return to the original level within 2–3 hr as is the case with normal thyroid function. Hypothyroidism may be associated with sexual precocity and inappropriate elevation of serum gonadotropins and prolactin. Replacement treatment with thyroid hormone corrects these levels.

Thyrotoxicosis is not an uncommon condition during adolescent development. Any interference that this may cause with the orderly process of secondary sexual development is slight. The diagnosis is generally easily ascertained by the previously mentioned laboratory methods. The T3 uptake and T4 are usually elevated. Infrequently the disease is due entirely to triiodothyronine secretion (T3 *not* T3 uptake) wherein T4 may be normal. Only the T3 is high. When employing the TRH test, there will be no response of the TSH as opposed to hypothyroidism. This latter test is rarely required to make the diagnosis.

A cautionary note is necessary. In young girls taking oral contraceptives, there will occur an elevation of the serum T4. There may also be a goiter present in a small percentage. However, there are no clinical signs of thyrotoxicosis. In this situation, the T3 uptake instead of being elevated as it should in thyrotoxicosis, is depressed. This is a consequence of the elevation of the thyroxine-binding globulin, which is produced by the exogenous female sex hormones.

So-called "simple goiter" of adolescence is fairly common in the United States. It is usually not attended by signs either of excessive or deficient thyroid hormone. However, it is not to be regarded as a normal phenomenon of this era. Most cases can be shown to be the result of chronic lymphocytic thyroiditis. This diagnosis can be ascertained by the determination of two antibodies in the serum, that against human thyroglobulin and the microsomal thyroid antigen. If one or the other of these is distinctly elevated, the diagnosis is fairly secure. There is an increasing tendency to employ fine needle biopsy in this condition, but this requires highly expert cytologic interpretation, which is not generally available at the present time. Open biopsy is not generally recommended unless the history reveals exposure of the neck and adjacent regions to irradiation in earlier life. This is a most important question to be elicited in the history and is sometimes not available directly from the patient. The importance of this condition lies in the high incidence of hypothyroidism in later life.

REFERENCES

1. Marshall WA, Tanner JM: Variations in pattern of pubertal changes in girls. *Arch Dis Child* 44:291–303, 1969
2. Zacharias L, Rand WM, Wurtman RJ: A prospective study of sexual development in American girls: The statistics of menarche. *Obstet Gynecol Surv* 31:325–337, 1976
3. Jenner MR, Kelch RP, Kaplan SA, et al: Hormonal changes in puberty. IV. Plasma estradiol, LH and FSH in pre-pubertal children, pubertal females and in precocious

puberty, premature thelarche, hypogonadism and in a child with a feminizing ovarian tumor. *J Clin Endocrinol Metab* 34:521–535, 1972

4. Winter JSD, Faiman C: Pituitary-gonadal relations in female children and adolescents. *Pediat Res* 7:948–953, 1973

5. Reiter EO, Root AW: Hormonal changes of adolescence. *Med Clin N Am* 59:1289–1304, 1975

6. Guida HJ, Friesen HG: Serum prolactin levels in humans from birth to adult levels. *Pediat Res* 7:534–540, 1973

7. Harayama N: The basal levels of serum gonadotropins and their responses to synthetic LH-RH in normal healthy children. *Acta Paed Jap* 17:57–58, 1975

8. Reiter EO, Root AW, Duckett GE: The response of pituitary gonadotropes to a constant infusion of luteinizing hormone-releasing hormone in normal pre-pubertal and pubertal children. *J Clin Endocrinol Metab* 43:400–411, 1978

9. Koenig MP, Zuppinger K, Liechti B: Hyperprolactinemia as a cause of delayed puberty. *J Clin Endocrinol Metab* 45:825–827, 1977

10. Bidlingmaier F, Wagner-Barnack M, Butenandt O, et al: Plasma estrogens in childhood and puberty under physiologic and pathologic conditions. *Pediat Res* 7:901–907, 1973

11. Hatch R, Rosenfield RL, Kim MH, et al: Hirsutism: implications, etiology and management. *Am J Obstet Gynecol* 40:815–830, 1981

12. Boon DA, Keenan RE, Slaunwhite WR, et al: Conjugated and unconjugated plasma androgens in normal children. *Pediat Res* 6:111–118, 1972

13. Apter D, Vihko R: Serum pregnenolone, progesterone, 17-hydroxyprogesterone, testosterone and 5α-dihydrotestosterone during female puberty. *J Clin Endocrinol Metab* 45:1039–1048, 1977

14. DePeretti E, Forest MG: Pattern of plasma dehydroepiandrosterone sulfate levels in humans from birth to adulthood. *J Clin Endocrinol Metab* 47:572–577, 1978

15. Kirschner MA, Zucker IR, Jespersen D: Idiopathic hirsutism: An ovarian abnormality. *N Eng J Med* 294:637–640, 1976

16. Migeon CJ, Rosenwaks Z, Lee PA, et al: The attenuated form of congenital adrenal hyperplasia as an allelic form of 21-hydroxylase deficiency. *J Clin Endocrinol Metab* 51:647–649, 1980

17. Bongiovanni AM: Acquired adrenal hyperplasia with special reference to 3β-hydroxysteroid dehydrogenase. *Fertil Steril* 35:599–608, 1981

18. Eisenberg E: Toward an understanding of reproductive function in anorexia nervosa. *Fertil Steril* 36:543–550, 1981

CHAPTER 11

Legal Considerations in the Treatment of Minors

Elisa Bongiovanni

1. INTRODUCTION

The practice of medicine has become intertwined with legal considerations as medical technology has continued to advance. New developments in science multiply both the human life span and the number and types of accompanying risks to the patient. These risks are one of the factors for the increase in lawsuits effecting physicians. However, it would be an oversimplification to view the popularity of medical malpractice cases in the last 10 years as contingent solely on technological advancements.

Health care has become a rather large systematized business. Insurance companies, hospitals, and continuing legislation all interact, and when changes occur too rapidly, the system as a whole often fails to assimilate the change and a breakdown occurs. Such a situation occurred in 1974 when malpractice premiums rose to an unprecented high. The reasons for this occurrence are rather circular. There is, of course, some evidence to indicate that the actual number of claims or lawsuits did rise. Insurance companies, however, threatened an exhorbitant increase in premiums, which neither individual physicians or hospitals were prepared to absorb. Hospitals became aware of their increased liability, and patients seemed to reflect an increased litigiousness, an attitude reflective of all society. All of these factors plus a rapidly changing technology worked to increase the number of malpractice suits.

This situation has necessitated the acquisition of baseline legal information by every practitioner. It is the intention of this chapter to familiarize the physician with those legal issues and concerns most frequently involved in the treatment of adolescents.

Elisa Bongiovanni • Thomas Jefferson University, Philadelphia, Pennsylvania 19107.

2. LEGAL BACKGROUND

All of law can be divided into two basic areas: criminal law and civil law or torts. Criminal law deals with crimes against society and actions are brought by the Commonwealth against individuals who are believed to have committed a crime. The purpose of the legal action in a criminal action is to both protect and vindicate the interests of society by punishing the offender and deterring others from imitating him.[1] The injured party is not compensated for the actions of the accused.

Civil law encompasses legal actions brought by one individual against another. The purpose of the action is to compensate the injured party for the damage he has suffered at the expense of the wrongdoer.[2] A successful civil action will result in a judgment of a specific sum on money.[3] Generally, civil law is the area of law that interacts with medicine, and it is the area of law with which physicians should be best acquainted.

Lawsuits against physicians for the type of treatment that they have rendered are labeled *malpractice actions*, in which the plaintiff or party who brings the action must prove that the physician was negligent in his actions. One definition of negligence is conduct that falls below a standard established by the law for the protection of others against unreasonable risk or harm. In a negligence claim, four elements must be proved. First, there must be a duty or obligation between the parties of such a nature that a particular standard of care is required. Second, there must be a failure on one party's part to conform to the standard required; that is, there must be a breach of duty. Third, there must be a reasonably close causal connection between the conduct of one party and the resulting injury of the other party. Lastly, there must be some actual loss or damage that has resulted from the above.

All of these preceding elements must be found to exist in order for a claim of negligence to be upheld in a court of law. While it is true that lawsuits that can be viewed as somewhat spurious are filed, it is also true that such cases generally do not succeed. Law, like medicine, is based on a system of interlocking blocks of knowledge. The wrong application of legal theory rarely succeeds.

There are other types of actions besides negligence actions in which a physician may be named party as a result of the care rendered to a patient. Basically, the remaining types of lawsuits are based on informed consent or lack of it. A subsequent section will address these issues.

3. SPECIAL LEGAL CONSIDERATIONS IN THE TREATMENT OF ADOLESCENTS

The term *adolescence* generally refers to youth who can be viewed as minors in the eyes of the law. Minors occupy a special status because of both societal and legal sanctions. Traditionally, the family has exercised dominion over matters relating to children.[4] This position was justified by the law as one way of keeping the best interests of the child in mind.[5] In instances where there has been a conflict between the parents and the states' views of what is in the best interests of the child, the supremacy of parental authority has been upheld. In *Wisconsin vs. Yoder*, 406 U.S. 205 (1972), the court held that Amish parents had a right to their religious convictions, which included withdrawing their children from school at an earlier age than that mandated by the state statute. *Pierce vs. Society of Sisters*, 268 U.S. 510 (1925), held that parents had a right to have their children taught in public schools, and *Meyer vs. Nebraska*, 262 U.S. 390 (1923), protected the parents desire to have their children taught a foreign language. However, while the preceding cases do give priority to parental authority, a new problem exists where the child and parent are not in agreement. Since the late 1960s, the courts have witnessed a rise in cases where not only was there a conflict between the parent and the state, but there existed the added conflict of the child. Perhaps the abortion issue best crystallized the competing interests of parent, child, and state.

The courts have to date skirted the issue of a minor's right to consent to abortion. The point of fact, the abortion issue itself, seems to be undergoing a change in this country at this writing. Nevertheless, in the leading test case of a minor's constitutional right to consent to abortion, *Bellotti vs. Baird*, 428 U.S. 132 (1975), despite efforts at forcing the court to decide on a minor's rights, no decision was made. The Justices themselves remained split on this topic. This remains a sensitive area, and the courts are reluctant to act quickly in this area of law.

It is obvious that support of parental authority in a situation where there is a clear conflict is unfair. Thus, when a minor seeks an abortion and a parent desires to prevent one, it is easy to see how parental authority could be abused. As one author has suggested, family privacy may become a cover for exploiting the inherent inequality between parent and child.[6] Since one of the prime functions of law is to prevent the dominance of any one group, it would appear inappropriate for the law to ignore the child's own interests in such a situation. The state has no independent interest in protecting the supremacy of parents.[7] However, this discussion serves only to explain some of the legal considerations faced by courts in expanding the rights of minors. Both society and parents have an interest in both protecting youth and allowing them expression of their constitutional rights. This area is indeed a troublesome one, and despite litigations, resolution of the issues has not been reached.

In point of fact, however, there has been a shift from judicial emancipation of minors to statutory emancipation.[8] This means that in many states, new laws are being developed that outline minors rights. This lessens the need for court hearings and determinations.[8] Furthermore, this shift gives some concrete direction to the practicing physician who treats adolescents. The following section outlines the most significant information pertaining to minors legal rights.

To summarize, most states do not allow minors to make contract obligations. However, almost every state allows minors access to some type of medical care. This means that a physician will be legally allowed to treat a minor but may not necessarily be able to collect payment for the treatment.

4. STATUTES SET THE LAW OF EACH STATE

Each state has a special set of statutes which encompass state law. Minors' rights to medical treatment are contained in these statutes. Generally, pregnancy, venereal disease, and drug abuse can all be treated with the consent of the minor alone. Many states also have a category labeled *emancipated minor*. This category sets up a criteria, which, if met, allows the minor to act as an adult in a certain limited and defined set of circumstances. Some of the items necessary for a minor to be considered emancipated are financial independence from parents, separate residence from parents, marriage, and parenthood. In order to determine whether a given state has such a category and what factors are used, a purview of the statute is necessary. Generally, emancipated minors may make all decisions relevant to medical treatment for themselves and any children which they may have.

Table I lists the states and outlines the basic rights each state gives to minors seeking medical treatment.

5. A PHYSICIAN'S LIABILITY FOR LACK OF CONSENT

A physician may be liable for battery if valid consent is not obtained. A battery is a tort or civil wrong and encompasses all physical contacts between people which are not consented to. Our society places high value on the inviolability of the individual, and, therefore, it treats such unauthorized contacts as a breach of legal right.

In the case of a minor, parents may bring suit against a physician for a battery. A minor will not have the legal capacity to commence a lawsuit. The physician, however, may erroneously rely on the consent of a minor for a certain procedure. A thorough understanding of the state statute where you practice will allow you to act with confidence in the various situations where consent is necessary. However, as a general rule, it is always preferable to obtain the parent's consent.

Some states, by special legislation, give added protection to a physician who believes he has received adequate consent from a minor. Pennsylvania, for example, in 1970, enacted a legislation that essentially states that if a physician relies on good faith upon the representations of a minor

TABLE I. Analysis of Minors' Legal Rights to Consent to Health Care Procedures in Each State[a]

State	Age of majority	Factors allowing those under the age of majority to consent to medical treatment	Any minor can consent to treatment of			
			Pregnancy	VD	Birth control	Alcohol
Alabama[9]	19	High school graduate, 14 years or older, or 14 years old and married, divorced, or pregnant	X	X		X
Alaska[10]	18	Married minors or minors living away from home and financially independent	X	X	X	
Arizona[11]	18	Emancipated minors and/or married minors		X		
Arkansas[12]	18	Married minors	X	X		
California[13]	18	Married minors, emancipated minors 15 years or older or minors on active duty with armed services	X		X	
Colorado[14]	18	Emancipated minors 15 years or older or married minors			X	
Connecticut[15]	18	Married minors or minors with children		X		
Delaware[16]	18	Married minor	X Must be over 12 years of age	X		
District of Columbia[17]	18		X	X	X	X
Florida[18]	18	Married minors	X		X	
Georgia[19]	18			X		
Hawaii[20]	18	Married minor	X Doctor must inform parent	X		X Counsel only

(*Continued*)

TABLE I. (*Continued*)

State	Age of majority	Factors allowing those under the age of majority to consent to medical treatment	Any minor can consent to treatment of			
			Pregnancy	VD	Birth control	Alcohol
Idaho[21]	18	Any minor 14 years or older can consent to hospital, medical, or surgical care related to treatment of infectious communicable diseases		X		
Illinois[22]	18	Minors who are married or pregnant				
Indiana[23]	18	Married or emancipated minors		X		
Iowa[24]	18 or marriage			X		X
Kansas[25]	18; 16 if married		X If no parent available	X		
Kentucky[26]	18; 21 if handicap	Emancipated or married minor	X	X	X	
Louisiana[27]	18	Married		X		X
Maine[28]	18			X		
Maryland[29]	18	Married minor and those who are parents	X	X	X	X
Massachusetts[30]	18	Emancipated minors, married, divorced, or widowed minor or minors with children	X	X	X	
Michigan[31]	18	Married, active military duty, or action of parents or courts		X		
Minnesota[32]	18	Emancipated minors or married minors	X	X	X	X
Mississippi[33]	21	Emancipated minors or unemancipated minors of sufficient intelligence		X		
Missouri[34]	21; 18 for medical care	Married minors or minors with children 18 years or older	X	X	X	X
Montana[35]	18	Emancipated minors, married minors, minors with children, or high school graduate	X	X	X	X

State	Age	Conditions				
Nebraska[36]	19	Married		X		
Nevada[37]	18	Emancipated minors, married minors, or mothers		X		
New Hampshire[38]	18			14 or >		X
New Jersey[39]	18	Married minors	X	X		
New Mexico[40]	18	Married minors or emancipated minors	X	X		
New York[41]	18	Married minors or minors with children		X		
North Carolina[42]	18	Emancipated minors	X	X	X	X
North Dakota[43]	18			X		X
Oklahoma[44]	18	Those active in armed services, emancipated minors, married minors, or minors with children	X	14 or >	X	14 or >
Oregon[45]	18	Married		X	X	
Pennsylvania[46]	18	High school graduate, married minors, or pregnant minors	X	X	X	
Rhode Island[47]	18	Married minors 16 years or older		X		
South Carolina[48]	18	Married minors 16 years or older	X	X	X	
South Dakota[49]	18			X		
Tennessee[50]	18			X	X	
Texas[51]	18	Married minors, emancipated minors 16 years or older, or active in military	X	X		
Utah[52]	18	Married minors	X	X		
Vermont[53]	18		X	X		12 or older
Virginia[54]	18	Married or divorced minors	X	12 or older	X	
Washington[55]	18	Minors married to 18-year-olds		X		14 or older
West Virginia[56]	18			14 or older		
Wisconsin[57]	18	Married minors		X		
Wyoming[58]	18			X		

[a] As new legislation is enacted continually, note that data in this table may not be current by date of publication.

regarding their ability to consent to treatment, this consent will be effective even though the minor is not one whose consent alone is legally effective.[59] This type of added protection grants the physician more comfort in his or her treatment of minors. Again, however, each state has a different set of rules and physicians must be familiar with the laws in their jurisdiction.[60]

It is suggested that physicians who regularly treat adolescents obtain a copy of the statute in their state that governs medical treatment of minors. This information can be translated into an office policy. In this manner, legal liability regarding consent issues will be minimized. This information can be obtained from a local attorney or possibly from the hospital where you are affiliated.

6. EMERGENCY TREATMENT

In any life-threatening situation, consent is implied and a physician is considered justified in assuming the patient's consent to treatment. However, this doctrine of "implied consent" applies only if the patient is not legally competent to give consent, and a parent or legal guardian is not present. The main possibility of legal problems arising out of emergency situations are in situations where a physician's interpretations of life-threatening is clearly divergent from that of his colleagues. From a legal perspective, an emergency situation is one in which a medical procedure is necessary for the diagnosis and/or treatment of a condition that if not treated immediately could lead to death or permanent disability.

In situations where a patient is being seen in a hospital facility, the physician should be sure to check hospital policy on dealing with these types of situations and should be sure to implement any policy of the facility.

An added protection can be obtained by careful record keeping. Physicians should be aware that medical records are the primary source of information once a dispute arises. Large gaps in a medical record unfortunately leave room for various interpretations. Clear, accurate, and thorough record keeping are an adjunct to the practice of good medicine. Accurate records also minimize the possibility of inferences that might not be true.

7. A PARENT'S REFUSAL TO CONSENT TO TREATMENT OF A MINOR CHILD

There are numerous reasons that parents or guardians may refuse treatment for a minor. However, if the procedure is deemed necessary for the preservation of life, a physician may decide to take some legal action to obtain authorization from the courts. Keep in mind that a physician who acts against the parent's expressed wishes and without judicial determination may be charged with battery. Hospitals and physicians may petition the courts for an order granting them permission to proceed despite the parents objections. Obtaining a court order provides the maximum amount of legal protection to the physician and the hospital. The courts, in reviewing such a petition, will consider such factors as the urgency of the child's condition, the benefits of the treatment, and the reasons for the parents refusal of the proposed treatment. In cases where there appears little dispute over the necessity of the treatment, the courts will rule favorably on the treatment. On the other hand, if the parents reasons for refusal are deemed good ones, the courts will uphold the parents mandate.

There are other situations where a minor either appears to be in physical or emotional danger due to the acts or behavior of a parent. In these cases, state law usually proscribes a method of action for physicians. In Pennsylvania, the Child Protective Services Law, Act 124, requires the reporting of all suspected abuse of neglect of a physical or psychological nature.[61] Additionally, the act sets up a mechanism whereby the state welfare department and the courts may remove a child from the parents either temporarily or permanently. This type of act not only provides a fast mechanism for the protection of a minor but is another mechanism for a physician to obtain medical treatment for a needy patient. Most states have statutes, and both the statute and the reporting form should be kept readily available. An abstract of the applicable Pennsylvania statute and reporting form are illustrated in Fig. 1.

8. COMPENSATION FOR SERVICES

Minors, by law, are generally not able to enter into binding contracts. A few states have by statute amended this rule of law, but, as a maximum, it holds true. Therefore, a physician who treats a minor without the consent of a parent may not be entitled to compensation.

In treating adolescents, it is, therefore, a good idea to speak directly with the patient about both the confidentiality of their communications with you and their method of payment. Adolescence is a difficult period, primarily, because of the transition from child to adult. A sensitive physician can show the patient in this category a willingness to recognize their new status by dealing with all aspects of medical treatment—this includes payment.

SECTION II. Immunity from Liability

"Any person, hospital, institution, school facility, or agency participating in good faith in the making of a report or testifying in any proceeding arising out of an instance of suspected child abuse . . . shall have immunity from any liability, civil or criminal, that might otherwise result by reason of such actions..."

SECTION 12. Penalties for Failure to Report

"Any person or official required (mandated) by this Act to report a case of suspected child abuse who willfully fails to do so shall be guilty of a summary offense, except that for a second or subsequent offense shall be guilty of a misdemeanor of the third degree."

24 HOUR TOLL-FREE HOTLINE: (800)932-0313

INSTRUCTIONS FOR COMPLETING THIS FORM

1. List the name, birthdate (or appropriate age if birthdate is unknown), and check appropriate sex. List the permanent address, zip code and county of child. Also, list present location of child, i.e., hospital, shelter, foster home, friend, relative if different than permanent address.
2. List the name, address and telephone number of natural/adoptive parent(s). Complete this section if information is known, even if child is not living with adoptive parent(s).
3. List person(s) responsible for child and with whom child lives, if information is different than that of natural/adoptive parent(s). Also indicate relationship to child, e.g., uncle, brother, parents, neighbor, etc. Indicate address and telephone number.
4. Indicate identifying information of person(s) allegedly responsible for the suspected abust and/or neglect of child. Indicate relationship to child, e.g., baby-sitter, teacher, day care mother, foster parent, mother, parents' paramour/boyfriend.
5. Give description of what actual injuries/neglect were sustained by the child. Include all reasons why child abuse is suspected, including how it occurred and any admissions of the act of abuse. Include any indications of prior abuse.
6. Check appropriate actions if applicable. Whenever any photographs or x-rays are taken, they along with any medical summaries shall be sent to the Child Protective Services of the Child Welfare Agency as soon as possible.
7. List the name, sex and age of each person in household under 18 years of age. Check block if there is any reason to believe that there has been any indications of prior abuse.

		Today's Date
1. Name of child (Last, First, Initial)	Birthdate	Sex M F
Address (Include Street, City, State and Zip Code)	County	
1A. Present Location If Different Than Above		

FIGURE 1. Protection of minors. The Pennsylvania statute: Sections from the "Child Protective Services Law," Act 124, November 26, 1975[62] and reporting form for suspected child abuse.

2. Natural/Adoptive Mother (Last, First, Initial)   Telephone No.

   Address (Include Street, City, State and Zip   County
   Code)

   Natural/Adoptive Father (Last, First, Initial)   Telephone No.

   Address (Include Street, City, State and Zip   County
   Code)

3. Other Person Responsible (Last, First, Ini-   Relationship to Child
   tial)

   Address (Include Street, City, State and Zip   County   Telephone No.
   Code)

4. Alleged Perpetrator (Last, First, Initial)   Relationship To Child

   Address (Include Street, City, State and Zip   County   Telephone No.
   Code)

5. Nature and Extent of Alleged Instances (Nar-
   rative)

6. Actions Taken or about to Be Taken

Notification of Coroner   Photographs   Emergency Custody Taken

X-Rays   Hospitalization   Other (Specify)

7. Family Household Composition
Name   Sex   Age   Name   Sex   Age
Name   Sex   Age   Name   Sex   Age
Name   Sex   Age   Name   Sex   Age

8. For Use by Physicians Only (Please Print or Type)
Name   Date   Hospital/Office

Medical Diagnosis: _____
                                                         Signature

9. Mandated Reporting Source (Other Than Physician)   (Please Print)
Name   Title   Agency

County   Signature _____   Date
10. Nonmandated Reporting Source
Name   Relationship to   Address (Street,   County   Telephone
                       Child   City, State, Zip
                                       Code)

FIGURE 1 (*Continued*).

9. CONCLUSION

A knowledge of law as it applies to the practice of medicine has become increasingly necessary as both the technology of medicine advances and as society becomes more litigous. Nevertheless, a knowledge of the law is a useful adjunct to the practice of good medicine. Adolescents present special legal problems, because, for the most part, they are under the age of majority and may not legally be able to provide adequate consent for medical procedures. The law, however, has recognized the unique and problematic status of adolescents and has in most states allowed these minors to consent to treatment for specific conditions. Each state will have different rules regarding this matter, and it is wise for every medical practitioner to become familiar with their state statute. The sensitive gynecological problems of pregnancy, birth control, venereal disease, and alcoholism are generally addressed in the state statutes, and more often than not, adolescents are given the legal latitude to consent to treatment for these conditions. As a concluding comment, remember that law like the practice of medicine changes with time. A periodic update of recent legislation in the above areas is necessary to stay abreast of new developments.

REFERENCES

1. Kenny CS: *Outlines of Criminal Law*, ed 15. Cambridge University Press, Cambridge, England, 1936, Chap 1
2. Prosser: *Handbook of the Law of Torts*, ed 4. West Publ, St Paul, Minnesota, 1971, p 7
3. Prosser: *Handbook of the Law of Torts*, ed 4. West Publ, St Paul, Minnesota, 1971, p 7
4. Previous decisions have recognized the importance of the family unit in our society. *Wisconsin vs Yoder*, 406 US 205 (1972); *Stanley vs Illinois*, 405 US 642 (1972); *Prince vs Massachusetts*, 321 US 158 (1944)
5. Kleinfeld AJ: The balance of power among infants, their parents, and the state, I–III. *Fam Law Quart* 4:320, 410 (1970, *Fam Law Quart* 5:64–107 (1971)
6. Goldstein J: Medical care for the child at risk: On state supervision of parental authority. *Yale Law J* 86:645, 661–662, 1972
7. Goldstein J: Medical care for the child at risk: On state supervision of parental authority. *Yale Law J* 86:663, 1972
8. Katz SN, Schroeder WA, Sidmon, LR: Emancipating our children—211 coming of legal. *Fam Law* 7:211–241, 1973
9. Code of Alabama: §26-1-1; §22-8-4; §22-8-5; §22-8-6; §22-8-3
10. Alaska Stat: §25.20.010; §25.20.020 (1978 Suppl); §09.65.100 (1979 Suppl)
11. Arizona Rev Stat Ann: §1-215 (1978–79 Cum Suppl); §44-132; §44-133 (1967 Bd. Vol.); §44-132.01; §44-133.01; §44-135 (1978–79 Cum Suppl)
12. Arkansas Stat Ann: §57-103 (1979 Cum Suppl); §82-363; §82-629 (1976 Repl Vol); §82-1606 (1979 Cum Suppl)

13. Cal Civ Code Ann (1979 Cum Suppl): §25.1; §25.6; §34.5; §34.6; §25.7; §34.7; §25.5
14. Colo Rev Stat 1973: §12-33-101; §12-22-103; §13-22-102; §13-22-105
15. Con Gen Stat Ann (1977 Ed.): §1-1(d); §19-89(a); §19-496(c); §19-142(a)
16. Del Code Ann (1974 Rev): tit 1, §701; tit 13, §707; tit 13, §709; tit 13, §708 (1978 Cum Suppl)
17. DC Code Ann: §21-101 (1978 Suppl); §6-119 j-1 (1973 Bd Vol); Regulation No 74-22, DC Register (Sept 16, 1974)
18. Florida Stat Ann (1979 Cum Suppl): §743.07; §743.01; §458.21; §743.06 §458,215; §381.382 (1973 Bd Vol)
19. Ga Code Ann: §74-104; §74-104.2 to 74-104.4 (1973 Rev); §84-932; §48-111; §88-2904 (1979 Rev)
20. Hawaii Rev Stat (1978 Suppl): §677-1; §577-25; §577A-1; §577A-2; §577A-3 §577A-4; §577-26
21. Idaho Code Ann: §32-101 (1977); §37-3102 (1976 Cum Suppl); §39-3801 (1977).
22. Ill Stat Ann: Ch 3, §131 (1978 Cum Suppl); Ch 111, §4501; Ch 111, §4502; Ch 111, §4503; Ch 111, §4504 (1978 Bd Vol); Ch 111–112, §600 (1977 Bd Vol)
23. Ind Stat Ann: §34-1-67-1 (1979 Cum Suppl); §16-8-4-1; §16-8-4-2; §16-8-5-1 (1973 Bd Vol); §16-13-6.1-23; §16-8-2-1 (1979 Cum Suppl)
24. Iowa Code Ann (1979–80 Cum Suppl): §599.1; §140.9; §125.33
25. Kansas Stat Ann: §38-101 (1978 Cum Suppl); §38-122; §38-123; §38-123b (1973 Bd Vol); §65-2891 (1978 Cum Suppl); §65-2892; 2892a (1973 Bd Vol); §38-123a (1978 Cum Suppl)
26. Ky Rev Stat: §2.105 (1971 Ed); §214.185 (1977 Repl Vol)
27. La Rev Stat Ann: Art 37 (1979 Cum Suppl); Art. 379; Art. 366 (1952 Bd Vol); §40:1095; §40:1065.1; §40:1096 (1977 Bd Vol); §40:1097 (1979 Cum Suppl)
28. Me Rev Stat Ann: tit 1; §73 (1979 Bd Vol); tit 32, §2595 (1978 Bd Vol); 22 §1823 (1977–78 Suppl)
29. Ann Code of Md: Rules of Procedure, Rule 5, §r (1977 Relp Vol); Art 43, §135; Art 43, §135A (1978 Cum Suppl)
30. Mass Gen Laws Ann: Ch 4, §7 (1979 Suppl); Ch 111, §117 (1975 Rev); Ch 112, §12F, Ch 112, §12s (1979 Suppl); Ch 111B, §10 (1975 Rev)
31. Mich Comp Law Ann: §722.1; §722.4; §701.19b; §722.124a (1979–80 Cum Suppl); §333.5257; §333.6121 (Pub Health Code, 1978 App)
32. Minn Stat Ann (1979 Cum Suppl): §645,451; §144.341; §144.342; §144.343; §144.344; §145.41
33. Miss Code Ann 1972: §1-3-27; §41-41-3; §41-41-7; §41-41-13; §41-41-15; §41-30-21; §41-42-7 (1978 Cum Suppl)
34. Mo Stat Ann: §475.010; §431.061; §431.063 (1979 Cum Suppl)
35. Mont Rev Code Ann (1978): §41-101; §41-1-402; §41-1-405; §41-1-406; §41-1-403; §41-1-405
36. Neb Rev Stat: §38-101 (Reissue 1978); §71-1121 (Reissue 1976; §71-4808 (1978 Cum Suppl) (See also §§44-2816, 44-1820, and 44-2821)
37. Nev Rev Stat: §129.010; §129.030; §129.040; §129.050; 129.060
38. NH Rev Stat Ann: §21-B:1; §318-B:12-a; §571-C:1; (1977 Suppl) §141.11-a (1977 Relp. Ed.)
39. NJ Stat Ann: (1976 Bd Vol): §9:17B-1; §9:17A-1; §9:17A-4; §9:17A-5
40. NM Stat Ann (1978 Repl Vol): §32-1-3; §24-1-13; §24-10-1; §26-2-14; §24-10-2
41. McKinney's (1977): Dom Rel Law §2; Pub Health Law §25-4; §3123; §2305; Ment Hyg Law §23.01
42. NC Gen Stat: §48A-2 (1976 Relp Vol); §90-21.1 (1975 Relp Vol); §90.21.1(4); §90-21.5; §90-21.4; §90-220.11 (1979 Cum Suppl)
43. ND Cent Code Ann (1977 Suppl): §14-10-01; §14-10-17; §14-10-17.1
44. Okla Stat Ann (1978–79 Cum Suppl): tit 63 §2601; tit 63 §2602; tit 63 §2604; tit 63 §2603; tit 63 §2152; tit 10 §170.1; tit 10 §170.2
45. Ore Rev Stat (1975–76 Repl Vol): §109.510; §109.520; §109.610; §475.742; §109.640; §109.650; §109.670

46. Pa Stat Ann Tit 35 (1977 Bd Vol): §10101; §10102; §10103; §10105; tit 71 §1690.112 (1979–80 Cum Suppl)
47. RI Gen Laws Ann: §15-12-1; §23-51-1 (1978 Cum Suppl); §23-11-11
48. SC Code Ann: §15-1-320; §32-5-30; §44-43-20 (1978 Cum Suppl); §44-45-30; §44-45-10; §44-45-20 (1976 Bd Vol)
49. SD Comp Laws Ann: §26-1-1 (1976 Rev); §34-23-16 (1977 Rev); §26-2-7 (1978 Suppl)
50. Tenn Code Ann: §1-313 (1978 Cum Suppl); §53-4607; §53-4401; §53-1104 (1977 Repl Vol 9B); §63-624 (1976 Repl Vol 11)
51. Tex Code Ann—Family Code: §11.01; §35-03; Texas Civ Stats art 4447i
52. Utah Code Ann: §15-2-1 (1979 Suppl): §78-14-5 (1977 Repl Vol); §26-6-39.1 (1976 Repl Vol)
53. Vt Stat Ann (1979 Cum Suppl): tit 1, §173, tit 18, §4226; tit 18, §9
54. Va Code Ann: §1-13.42 (1979 Repl Vol); §32-137 (1978 Cum Suppl)
55. Rev Code Wash Ann: §26.28.123; §26.28.020; §69.54.060 (1978 Suppl); §69.54.070; §70.24.110; §70.96A.110; §70.96A.120 (1975 Bd Vol)
56. W Va Code Ann: §2-3-1; §16-4-10 (1979 Repl Vol); §60A-5-504 (1977 Repl Vol); §60-6-23 (1979 Cum Suppl)
57. Wis Stat (1979–80 Suppl): §990.01(3); §49.10(6); §143.07; §51.13; §48.373 (1979 Bd Vol)
58. Wyo Stat Ann (1977 Repub Ed): §14-1-101; §35-4-131
59. 35 PS §10105
60. Rosoff, AJ: *Informed Consent: A Guide for Health Care Providers* Aspen Publ, Rockville, Maryland, 1981
61. Child Protective Services Law, Act 124, PL. 438, November 26, 1975

RECOMMENDED READING

Boli-Bennett J, Meyer JW: The technology of childhood and the state: Rules distinguishing children in national constitutions 1870–1970. *Am Sociol Rev* 43:797–812, 1971
Child Protective Services Law, Act 124 P.L. 438, §1, November 26, 1975
Fertility and Contraception in the United States, Report of Committee on Population, U.S. House of Representatives, 95th Congress, Second Session, p 66, December 1978
Pipel: Minor's rights to medical care. *Ala Law Rev* 36:462, 1972
Siliciano, JA: The minors right of privacy: Limitations on state action after Danforth and Carey. *Col Law Rev* 77:1216, 1977

Index